ADVANCED, EVOLVING WISDOM BEYOND SAPIENS´ MENTAL CONSTRAINTS, ESCAPISM AND SELF-DESTRUCTION

ADVANCED, EVOLVING WISDOM BEYOND SAPIENS´ MENTAL CONSTRAINTS, ESCAPISM AND SELF-DESTRUCTION

BENJAMIN KATZ

To order additional copies of this book, contact:
Xlibris
844-714-8691
www.Xlibris.com
Orders@Xlibris.com
839047

CONTENTS

Tikkun Olam/Adam

Aiming to enjoy life as a purpose in itself is a task for self-indulgent children, infantile hedonists, and narcissists, not for a fully developed man. He'd rather try to take care of his mental/physical health while dealing with Tikkun Olam/Adam (the improvement of man and our world). Maybe this is what keeps the mind young, vibrant, and aspiring for the stars.

The best deal I can strive for in my life is to have proactive ideals, a viable future mission/vision to pursue, and a reality that challenges me to work together with others for higher goals than my own petty life, while engaging myself in them in order to better/improve both man and his world by unfolding my best potentials.

My future vision is to:

Invest heavily and ceaselessly in upgrading many sapiens—as to be humanity fire column—to become much wiser, farsighted, and cooperative than we are today; in order to create a sustainable, ever evolving, just, and fair civilization that will head to the stars, be free of greed, rich people, as well as human degradation and exploitation.

It promotes cutting down drastically over the coming centuries on global population, global consumption, global pollution; abolishing global consumerism ideology (capitalism); reducing human shortsightedness, divisive propensity, excessive self-interest, context-free convictions, and megalomania.

Will people listen to my message?

They will accept this vision in due time as long as it is free of fantasy, lunacy, or profanity!

All these three we already have in our time: the reign of shortsightedness, self- deception, and greed. They will first stale but latest in this century when the global predicaments will become unbearable. First the whip will hit the donkey's back hard, and then it will move...

Preface

I proclaim:
What is the difference between you and me?
You are twenty, I'm soon eighty,
four times as old as you are.
You are already immersed in life's replications
I am engaged in grand life's exhilaration,
challenging its premises without hesitation.

You surrender to your soul's erosion
and to your free will corrosion,
trying to avoid your inner implosion.
while you nurture a global explosion.

You suffer of pleasure-seeking obsession
denying thereby your mindset's repression,
turning our ultimate meaning into diversion.
There will come a time to take a stand
not allowing fools to get the upper hand.
Where we'll sacrifice our lifestyle's race
In order to prevent our species' death!

What is wrong with us and with our civilization that we destroy our life conditions on Earth? We suffer from lack of foresight, of prudence, of measure in all, and above all transcending/advanced, evolving wisdom.

Seemingly enlightened people who until recently thought of the climate crisis as a kind of barter between us and our descendants—what we are willing to pay for our quality of life today so that they can live better in the future—now understand that this is not the case. The crisis will not only turn the lives of our children and grandchildren into hell, it is already here. It's going to be awful, and it's going to happen in our shift on earth.

The payback is coming: Many bloodier conflicts and devastating natural catastrophes; massive waves of refugees; barred borders and desperation, hunger, and poverty; deadly epidemics; and mass death in this century and beyond.

It becomes apparent that our brains are not designed to neither avoid nor resolve such a global catastrophe. Therefore, we need to upgrade ourselves to become much wiser/farsighted than the common sapiens are. We need to stop brainwashing children and grownups to act semiautomatically, whether it is a matter of religion, ideology, ethnicity, or hedonism/consumerism, and to work instead for a united, sustainable, evolving global civilization.

But in order to come so far, we must look the enemy straight in the eyes, and the enemy is our own limited mental power. We still think/act in many respects as animals (uninhibited multiplication, dominance/submission, wars/conflicts, impulsive behavior, and shortsightedness).

We know by now that many of our sophisticated political/economic models are not always good constructions to deal with our murky sides since they often don't include human irrational behavior, our propensity for over- optimism, greed, and irrational convictions.

Since the human brain has not changed much since the Stone Age—our brains are, in fact, shrinking—it becomes obvious that we have to give it a qualitative booster away from our beastly inheritance.

My vision has not yet won many disciples, clearly enough, as it rejects strongly our materialistic golden calf obsessive dance, and deals with our upgrading, aimed at making us much wiser than we are today, and demands self-sacrifice for the sake of reestablishing global sustainability. It will, nevertheless, impact the surviving civilization in due time, already in the end of this century, due to impending necessity. My vision will come to impact future civilization, not by seduction but by the stick/carrot as humanity's growing tribulations, despair, and necessity will force it to choose the direction it points out.

Working on this book, I have identified forty-six critical areas where our folly manifests itself to undermine our free will, and future prospects as well.

My view of this advanced, evolving vision/mission/wisdom is based on attaining four main general goals:

1) Self-knowledge resulting in improvement and acceptance of oneself, and as self, correcting/containing defense against human impulsive and shortsighted behavior

2) Knowing the world and reality beyond stiff convictions and dogmatic religions/ideologies, resulting in taking care of the planet's life conditions in a sustainable, global manner

3) Working for the betterment of both humans and the world (further evolvement and purifying the noble within us on account of the beastly part)

4) Pursuing our ultimate meaning, which is to become creators in our own right through our deliberate and conscious evolvement beyond sapiens and Mother Earth/our sun system/galaxy.

In order to attain such lofty vision/mission/wisdom, this book focuses on 1) identifying the areas where our fatal folly is manifested, and 2) on the proper course of strategies and actions to follow that can grant our progenies a promising future.

Our ongoing, modern, Chelmer's story

"Basically, most humans are slaves of their convictions, perceptions, pretense, and self-pretense. Only by fighting them head-on can we become momentarily free before we fall again in a new trap of this cocktail, and therefore the struggle must go on until we reach a more advanced brain/consciousness" (B. K.).

"The real wise among us don't tread the moral high ground as shifting contexts and farsightedness is necessary in order to deal with the complexity of reality and us" (B. K.).

On the day Kabul fell into the hands of the Taliban (15.8.2021) US Secretary of State Anthony Blinked insisted that the US had succeeded in its mission in Afghanistan, despite the Taliban takeover, adding that the US had no interest in staying there. He said the United States had invested billions of dollars in training Afghan forces to give them an advantage over the Taliban, but that they had failed to repel the terrorist organization. The Taliban` takeover happened faster than the American expected, he added.

Around the same time, the Atlantic Ocean currents seemed to be weakening, nearing the verge of collapse, a study said.

Imagine a world where North America is locked in snow. Winter storms ravage Europe, while Australia bakes in permanent drought. This isn't just the plot of Roland Emmerich's 2004 disaster flick "The Day After Tomorrow." It's also what could happen if a crucial network of currents in the Atlantic Ocean were to shut down.

Unfortunately, research published Aug. 5, 2021 in Nature Climate Change suggests that these currents are weakening due to manmade climate change. And if nothing is done to prevent it, they may collapse completely.

These two events indicate that human Chelm's reign has become global.

Once upon a time, there was a town called Chelm. In Chelm lived Chelmer, all endowed with a certain kind of well willing, good-natured foolishness. Rumors said that the town's inhabitants did not suffer too much of their legendary foolishness, as they were both poor and not so many, so they could not do much harm to their environment.

We have in our possession some observations demonstrating the state of a global Chelmer from old times, telling the enervating, often fruitless struggle of the few wise ones to sober up the countless, rapidly multiplying Chelmer of the world.

Chelm existed long before our time (human stupidity), but as time passed the Chelmer's foolishness spread slowly around the world, while mutating into a desire to become both rich, spendthrift, greedy, and multiply uncontrollably.

Of course people did not realize that they were infected by the Chelm virus, as fools don't realize being fools. But slowly their misdeeds had such drastic consequences that they could not fix, and so the perceived—false—wisdom they attributed to themselves began to crack.

What were the consequences that shook their trust in their unwavering wisdom?

Summer 2021 had just begun, and from late June to the days of mid-July the climate crisis was repeatedly giving its signals, with new records breaking all over the world. Peak temperatures under "outdoor heat" in Canada and North America had risen to nearly 50 °Celsius, and June was recorded as the hottest in history. A few days later, a heavy storm swept through the western United States, and the temperature measured in the Death Valley in California—the lowest, driest, and hottest area in the United States and one of the warmest on Earth—was 54.4 °Celsius, the highest temperature ever measured in the area. All of these resulted in many people killed, huge fires, significant damage to infrastructure, and the destruction of homes. Another report presented a frightening finding, according to which the heat wave in North America resulted in the deaths of millions of marine creatures.

Peak temperatures were also measured on the other side of the world. In the city of Jacobabad in Pakistan, a heat load was measured that combined high temperature and humidity equivalent to 52 °Celsius. And in the city of Jahar in Kuwait, a record temperature of 53.5 °Celsius was measured.

Along with the heat waves, strong storms also appeared from east to west. In Japan, in the town of Atomi near Tokyo, 313 mm of rain fell in twenty-four hours, above the average for the entire month of July (242 mm). And although heavy rains, floods, and mudslides are events in the rainy season routine, it was an extreme event.

Hurricane Elsa hit the East Coast of the United States and the Caribbean, with heavy rains, strong winds, and flooding, including flooding some of New

York's subway stations. Especially in Germany and Belgium, heavy rains fell and rivers overflowed, sweeping hundreds of people to their deaths and sowing much destruction in sights not seen before.

The melting of the Parma frozen area in Siberia, which is 11 million square kilometers, has been dramatically accelerated in comparison with models, which anticipated the current change to happen seventy years from now, and vast areas around the Arctic are reaching a tipping point (Climate Panel UN report, September 8, 2021). All the while China—only one country out of many, which is the biggest CO_2 polluter, producing more than a quarter of the world's CO_2 emissions—refuses to follow the FN CO_2 Curbing quota proposal.

All this havoc was caused by the wisdom of countless good-natured and decent and less decent Chelmer.

Modern Chelm has bestowed humans with power and facilities, but it has snatched from them golden rules like prudence, measure in all, humbleness, justice, compassion, and empathy regarding the coming generations. Postmodern Chelm has failed in building strong human character. It has corroded souls with the corrosive effects of materialism, selfishness, superficiality, and hypocrisy. It has given us wings to fly through spacious atmosphere, but have cut our roots from the very depth of strong character.

Modern Chelm' stupidity has caused the deterioration of many noble human virtues. Positive realism has been replaced by greed and overindulgence. Modesty by arrogance. Simplicity by glamour and character ethics has been reduced. Cooperation and integrity were substituted by negative competition. Authenticity by superficiality. Spirituality has been buried under the debris of negative rationality. Pollution and massive self-indulgence prone brain washing have dumbed down the populace. To cut it short, humans became worse humans regarding their virtues and their responsibility to the coming generations.

Environmental Crime in Modern Chelm:

This environmental crime has displaced or killed untold numbers of people around the world, caused billions of dollars in economic damage and ravaged vital ecosystems and wildlife. It has disproportionately affected already marginalized communities around the world, from farmers in coastal Bangladesh, where the fast-rising seas are salting the soil and slashing rice yields, to low-income residents of Houston, Chicago, and other cities, whose neighborhoods suffer higher temperatures than prosperous areas across town.

This crime threatens today's young people most of all, and calls into question the very survival of human civilization. And yet the criminals responsible for this devastation are still at large. Indeed, they continue to perpetrate their crime and make big money from it, not least because their crime remains unknown to most of the public.

One of the culprits of this crime is the fossil fuel industry's forty years of lying about climate change. Arguably the most consequential corporate deception in history, the industry's lies have had the effect of blunting public awareness and governmental action against what scientists say is now a full-fledged climate emergency.

Chelm's political behavior:

If you are hell sure that your society has a patent on the right way of living a good life in freedom, you are brainwashed.

USA and the west's terrible defeats in their invasions in this century are due to their naivety, Chelm's idealism (that democracy will be the chosen political system in the world).

Since the Berlin Wall's fall and the disintegration of the Soviet Union, most people in the west imagined that now the world will become democratic. In the 2000s, after 9/11 the west meant at for at promote this goal we had to use some force in other countries. In the 2010s, the west began to realize that this method does not work very well, and now in 2020s we in the west have given up the idea of democratic/liberal world and are building walls around us. In 1989 there were around fifteen walls and fences. Today there almost 100 of them.

Our defeats unleashed huge waves of Muslim immigrants/refugees in Europe, threatening to change countries own national identity.

Sweden, which was a well-functioning society in the sixties to the nineties in the twentieth century, took in many people from these areas. And the result: a society that has changed very much for the worse.

Most immigrants and refugees come from areas in the world that make it hard for people to survive due to overpopulation, a backward economy, criminality and political oppression/chaos, and not the least, climate change, which has turned big chunks of land uninhabitable.

How many countries are involved in conflicts and warfare in Chelm's world in 2021?

There are many, starting with the big powers (China, USA, Russia), India, Pakistan, Afghanistan, Iran, the Arabic Sunni world, Yemen, Libya, Syria, Iraq, Israel, Lebanon, Gaza, NATO, many countries in Africa and around China... and the list is long.

Another part of the Chelm's story is the deterioration of the population's mental/physical well-being.

Mental health problems increasing in global Chelm

Gullible people are always at risk. Cynical and decadent culture that encourages materialism/vain pursuit of narcissistic goals over real substance instead of purpose, determination, and pursuit of a great mission will make their citizens self-focused and mentally fragile. Lacking global, national, or social evolving schemes and great purpose in life results in many people floundering in their personal lives.

The Chelm population's mental health is deteriorating badly worldwide. There has been a 13 percent rise in mental health conditions and in substance-use disorders in the last decade (up to 2017). Mental health conditions now cause 1 in 5 years lived with disability. Dumb down of whole populations are taking place due to pollution for air.

Rates of mood disorders and suicide-related outcomes have increased significantly among adolescents and young adults, and the rise of social media may be partly to blame. Mental health problems are on the rise among adolescents and young adults. And the hedonistic/self-centered lifestyle, meaninglessness beside focus on consumption, career, and social media may be the driving forces behind this alarming increase (19 March 2019).

Obesity is on the rise in global Chelm:

Worldwide obesity has nearly tripled since 1975. In 2016, more than 1.9 billion adults eighteen years and older were overweight. Of these over 650 million were obese (9 June 2021).

Obesity is linked to both the shrinkage of our brain and to diabetes and some cancer types. It is an alarming development: while the average sapiens' body volume increases, his brain seems to decrease.

Obesity is a significant risk factor for and contributor to increased morbidity and mortality, most importantly from cardiovascular disease (CVD) and diabetes, but also from cancer and chronic diseases, including osteoarthritis, liver and kidney disease, sleep apnea, and depression (1 November 2010). This epidemic is primarily driven by rapid urbanization, nutrition transition, and increasingly sedentary lifestyles.

Summing it up:

We, well-meaning Chelmer, have created an unsustainable humanity, which destroys life conditions on Earth, and at the same time creates more mentally sick people, physically sick people, and more failing countries (due to lack of resources and too many inhabitants, and the immigration of often incompatible, culturally, people to well-functioning societies, thereby undermining them). Many conflicts keep flaring up all around the world. In chapters 1 and 2 I will try to illuminate how our Chelmer genius is our worst foe! In chapters 3 and 4 I will try to show a way out of this Chelm trap of our minds and behavior.

Chelm: From Yiddish literature/folklore, "The 'legendary' town inhabited by befuddled, stupid, foolish, but endearing people."

Introduction

My life's course, verdict on humanity, and its future direction Hitch our wagon to the stars

Moses was God's mind and hand
leading his folk to a Promised Land,
Yet he was buried in Sinai's sand
but his evolving story will never end...
—B. K.

"My mind is planted on the ground. My heart in the deepest human quest, while my soul is invested in transforming us into wiser and farsighted intelligent beings" (B. K.).

My life's destiny?

Look at this seventy-six-year-old man. You see me, but not really me deep inside. You see a sapiens. Yes, I am a sapiens, like you, but I wish to transcend this debilitating phase of my development, as I feel myself put in a mental straightjacket. Why so? Because just as all human beings, I am partly a beast, consisting of too little star dust and too much mud. I am prone to be shortsighted, self-glorifying, greedy and self-focused, pathetic, self-deceptive, and a great denier of facts on account of untested conviction. I act often automatically, semiautomatically, destructively and self-destructively, and hide these facts from my awareness through cognitive dissonance mental programs. All these mental programs are ingrained in my brain, and self-adulation and unfounded hopes camouflage them from my scrutinizing mind. I must come out of this straightjacket before it destroys me.

I was granted what most people would consider a very rewarding, lucky life. I was born in October 1945, shortly after the two atomic bombs were cast on Hiroshima and Nagasaki. When I look back at my participation in three wars (two of them as an active officer) and on the physically/mentally hard life I experienced in my kibbutz and military service in Israel, I consider myself lucky, as I came out of these epochs hardened, robust, empathic, and probably also wiser, without sores on either my body or soul. I never had a mental breakdown or depression. My body serves me pretty well at the age of seventy-six, without pains or illness; while my mind, being balanced and pro-active, sprouts ideas and exciting insights regarding the human condition/nature and our possible transcending future all the time.

I have had success with both friendship, love, and family, and have been free from addictions, obsessions, and compulsions that tyrannize so many sapiens in the modern world.

I have helped more than thirty-five thousand people over the years as a psychologist by teaching them to deal with problems and conflicts in their lives and by infusing them with courage, determination and life meaning.

Since 1998 I have written no less than 14 books (including this one)- mainly in English, but also in Hebrew and Danish- on psychology, on my future global vision and also some fiction.

I wrote much on the oncoming disastrous global climate change long before people came to consider it as a threat. On the *Titanic* syndrome, which seems entrenched in human minds: having fun "on the deck" while being oblivious to the oncoming danger. Eight billion consumers on earth are consuming/polluting unsustainably, thereby destroying their life conditions, believing that their seductive life party can go on unchallenged by changing to green energy, while the wrath of climate collapse is already hitting us hard.

While growing up and maturing, some kind of restlessness/hitch to the star's aspiration and melancholy came to take hold of me. I did not understand why such a lucky chap like me could not be thankful, content, and appreciative of his good fortune and just conform and be content. But I was only partly thankful, as in the back of my mind I had a premonition of oncoming global disaster due to humans' shortsightedness, and the need to avert it.

I noticed that I have possessed throughout my grown-up life an unsated appetite for exploring new vistas, horizons, and out of conventional box´ ideas both out in the world/universe and regarding the very human nature/condition. I was/am per se an explorer.

Next, I realized that my life motto has always included a needed and protracted struggle in order to achieve lofty goals in life. I was never a true peacenik, although I love peace with people. I was/am a warrior, believing in the dictum that without a real struggle in life to achieve something greater and more life affirming than the existing, we betray our potential mission in life: bettering humans and their world.

And lastly my driving forces turned out to be a trinity as I also became a visionary by writing a global vision. In my books I presented a global vision/mission for future human civilization away from sapiens' mental constraints, escapism, and self-destructiveness.

In my book *A Paradigm for a New Civilization*, I pointed out four cardinal goals: global sustainability, new global moral codex and conduct, upgrading of sapiens to become much wiser and farsighted than we are today, and to reach the stars with difficulties. Why to the stars with difficulties? I meant that as the restless explorers we are we need some grand challenges and struggles to keep ennobling and evolving ourselves. I used also a metaphor in my books *To Get Back as the Enliven Dust We Are* and *To Our Cradle*: the stars. I read in *New Scientist* ("Across the Universe," May 8, 2017, p. 1) that around 50 percent of the atoms in our bodies are from another galaxy. Can it be that those of them in my brain talked loudly when I devised my new vision?!

I belong to a miniscule group consisting of people who are both explorers, warriors, and visionaries. Their restlessness, their ever-going struggle to realize *Tikun Aolam Vaadam* (the bettering of both the world and ourselves), and their drive to put forth a vision is their hallmark. I am 4,000 years old, knowing the history of my people, plus 1,000 years old, peering into the future mists, anticipating and hopefully forming the journey of our progenies equipped with advanced, evolving wisdom.

Being entirely down-to-earth—concrete—or entirely spiritual blocks the potentiality of becoming a bit wiser. Without a viable vision your life has no mission. Shrinking one's universe to a comfortable size means becoming partially blinded.

My father had said once that the most tiring people in the world are those who are constantly preoccupied by their *pupik* (Yiddish, sexual organs) meaning their sole interest/focus is their ego, pleasure, and personal preferences. Being a disciple of Marxism most of his life, this point of view resonated well in his life. What Danish people lack—probably also other people and nations—is the aspiration for something greater than their *pupik*. This self-focus, enhanced in movies by professionals who are supposed to be the opposite of what they became, is tiring me with its endless replications and lack of greater horizons and perspectives.

My vision is in a state of a deactivated implant in the human collective Brain, waiting for the great collective sufferings to come in order to be activated...

I know for sure that my vision will live and be realized after I am gone. I am not going to go into its promised land, like Moses, but what the heck if my vanity will be frustrated, if it is going to lead to a much better one for the coming generation.

Verdict on current humanity

Sapiens' humanity is debilitating, shortsighted, and is arrested by life's actuality and banality's mental straightjacket. The most of us cannot think ten years from now, even though we have the scientific capacity to peer into the future 100 trillion years from now.

You may take a little trip through the universe in the next 100 trillion years just to understand that thinking, as I do. A thousand years ahead is no big deal. This trip may enlarge your overview of the coming world. (What will happen to our planet and our universe over the span of 100 trillion years? Quite a lot actually. Watch the video to find out...)

Humanity consists of a multitude of humans focusing their attention on their ephemeral significance while the world in which they live becomes a hostile place to live in. Their efforts to save their world and its habitats are too little and too late. Why is it so? It is because most of the people on earth lack proactive self-knowledge and the knowledge of their real reality, free of their convictions and faith.

Real reality consists of natural laws, where there is a food chain and we are on the top of it, and we eat animals and they eat each other, without any consideration of a merciful God. The real reality means human beings with much mud and a little bit of star dust: complex, contradictory, noble and debased, greedy and compassionate, a mixture of contradictions. These human beings strive for heaven but can easily be deluded by their sort of man's dementia originated from illusions/delusions, untested convictions, boasting, and self-glorification.

Real reality is a very muddy reality that most of us try to escape its seeming meaninglessness by granting our lives a transcending meaning. Most people have dubious relation to hard facts; either they ignore them or bend them as it suits them and their convictions. The poet T. S. Eliot said: "Humankind cannot bear very much reality."

Rejection of reality transforms many into publicly accepted psychotics.

No self-knowledge = No advanced, evolving wisdom!

People can be incredibly astute about the shortcomings of their friends, spouses, children, clients, and other people. But they often lack self- insight. The same people who are coldly clear-eyed about the world around them have nothing but fantasies about themselves. Self-knowledge doesn't work if you look in a mirror. This bizarre fact is, as far as I know, universal. Why so?

Personally, I have for a long time thought that there is a parallel for this problem visible in the computer world, in a procedure called recursion. Recursion means making the program loop back in on itself, using its own information to do things over and over until it gets a result. You use recursion for certain data-sorting algorithms and things like that. But it's got to be done carefully or you risk having the machine fall into what is called an infinite regress. It's the programming equivalent of those funhouse mirrors that reflect mirrors, and mirrors, ever smaller and smaller, stretching away to infinity. The program keeps going, repeating and repeating, but nothing happens. The machine hangs.

I have figured out that something similar must happen when people turn their defect self-knowledge on themselves. The brain hangs. The thought process goes and goes, but it doesn't get anywhere. It must be something like that, because we know that people can think about themselves indefinitely. Some people think of little else. Yet most people don't seem to change as a result of their intensive introspection. They don't understand themselves better. It's very rare to find genuine self-knowledge.

You need someone else to tell you who you are or to hold up the mirror for you. Once you possess genuine, proactive self-knowledge you can move toward acquiring advanced, evolving wisdom to guide us.

Advanced, evolving wisdom implies deep understanding of human nature/condition and reality beyond human constructs/convictions, and possessing sustainable, socially just, and evolving vision aimed at our upgrading and our future journey as creators. Advanced, evolving wisdom means a proactive struggle for intelligent life as such, to break away for our mental/physical "straightjacket" constraints through all available and future technologies. It advocates big reduction in global population, consumption, pollution, and production, and ceaselessly challenging our mind's/brain's constraints.

Our minds become mentally clogged/besieged when we have played out all our good cards without winning what we've dreamed of. This is the state of our civilization!

We face an unprecedented challenge/danger in human history: in order to survive, transform, and evolve on a long-term basis, we must reduce our global population (to a level of a maximum four billion), production, pollution, material greed, our shortsighted minds and mindsets, and distribute Earth's resources fairly to only those who follow this strategy/vision.

Not reducing global population, consumption, pollution, and growth ideology/mindset, and not becoming much wiser than we are today is a death kiss for our civilization.

My future direction: My vision

The direction of my long, winding road
does not rely on Almighty God,
neither on earthly, pretentious saviors,
but on sustainable, evolving warriors...

My direction points toward nurturing people who will become the fire columns of the new vision, who would go in front of us, guiding us on how to prevail by transcending.

This idea of evolving beyond our mental limitations will move on, fertilizing the coming future's people long after my body/mind has been dissolved. This spirit lived before I came to exist, settled down in me and others, and will move on, as it is the only real force we can reckon with to challenge and defy the rules of determinism and chaos. In this sense I am mortal but also immortal.

I don't really care regarding my popularity today, but I hope/expect that my ideas/vision will be useful from around 2050!

Hitching my wagon to the stars is what makes me and others immortal. And here and now we start our hitching with my wagon (vision) to the stars.

Chapter 1

WHY HAVE OUR MINDSETS/
MENTAL LIMITATIONS BECOME
OUR MORTAL THREAT?

"It is obvious that our most formidable yet invisible foe is our sapiens brain, which makes us blind to our beastly, fatal shortcomings" (B. K.).

"Human folly has always been
our biggest source of sins.
With it we follow a blind way,
leading our evolvement astray.
How can ourselves we save,
an evolving future to pave?"
—B. K.

"It is for most of us an invisible, frightening mental defect that we suffer of: shortsightedness combined with greed, which makes us prioritize our material privileges even when we can eye the oncoming deluge" (B. K.).

"The problem with the world is that the intelligent people are full of doubts, while the stupid ones are full of confidence" (Charles Bukowski).

He, Charles Bukowski, was only right regarding stupid people who are full of confidence. Intelligent people equipped with emotional balance/solidity can doubt but not become paralyzed by it, making a decision and pursuing it. Intelligent people full of speculation, deliberation, and self-doubt can't make up their minds. There are also the wise people—a minority among us—who know exactly what to do in this confused world, but they are not listened to. This is the problem of the world.

All my life, I have seen people who have controlled, denied, negated, sacrificed, suppressed, disciplined, and tortured themselves. And I say, for what? For God, for truth? A mind that has been tortured/massively brainwashed is crooked, rigid. Can such a mind see clearly reality? Certainly not.

The state of man:

Our Stone Age brain is hardwired to the following mental programs: God fixation, greed fixation, golem unaware programs (rituals, replications, automatization), group-think fixation, and immense gullibility. You can spend your life asking why we got such a difficult brain to manage our lives with, yet the answer is blowing in the wind.

Our repertoire regarding managing our lives is not impressive, and consists of five possibilities:

1) Being actively self-destructive/destructive, which many people pursue
2) Being resigned, passive, and giving up
3) Being ritualized/compulsive (the more ritualized/compulsive a human becomes, the more he distances himself from knowing himself and his world)
4) Being greedy, comfort and pleasure seeking
5) Working for a better world/human beings and struggling life-affirmingly to attain this goal

The state of our civilization:

Modern civilization faces three immense crises: ecological (unsustainability, pollution, dwindling resources and growing population and consumption), growing political and moral chaos, and growing mental/psychological disturbances in the global village.

The political global semi-chaotic state of today means that the two other crises (ecological and mental) can't be tackled effectively, as countries can't coordinate their efforts.

The western way of thinking since World War II and the Cold War was based on two sides. This sentiment, which always sees two sides in the global geopolitical picture (from this side, and from the other side), continues to exist, but it has completely lost its validity. Pretty quickly it turned out that this was not an equation at all, just an empty password. The United States is a superpower with unprecedented forces in history, yet such an image vis-à-vis Russia and China, which is indeed an emerging superpower, shows a multipolar reality. The truth is that the global reality today is of many sides. There is Russia, China, the EU, the US, as well as the rise of new forces in Africa and Central and South America.

The question is not whether a multipolar world arose but what its model will be. A world of globalization, where small and medium-sized countries have the same rights and status as big ones, as we have seen so far, or a world with many poles in the Russian or Chinese style. The Sino-Russian meaning is that there are agreed "areas of influence" between the strong countries, and the weaker countries are subject to their wishes.

The private identity of human beings has also changed dramatically, and now consists of ethnic, religious, national, gender, intellectual circles, and occupational/lifestyle. They are much more diverse in an age of globalization than ever before. A person can live on one continent, work in a society that operates on another continent, and have extensive social connections in the online world with people living in a different place. If our personal identities become more complex/fragile/fluid and the world has become multipolar, what can we expect from civilization?

What is required in this reality is to strengthen international bodies that can monitor and even establish laws in a global reality that seems increasingly chaotic. States do not want this central order because they fear the weakening of their sovereignty. This is our time paradox: We need global cooperation more than ever—our survival depends on it—yet some forces push this cooperation aside.

Take for example climate change, too many people, too much pollution, production, and consumption. Altogether, when not dealt with globally, will mean the end of our civilization.

The Australian climate researcher Will Steffen is an expert in the so-called tipping points. He asserts that when climate change reaches a certain point it will be too late to stop this development. Let us say that at the current pace of the transition to sustainable production and way of life we have become CO_2 free in forty to sixty years. It sounds good, but not good enough, because by then the ice at the North Pole, the permafrost in Siberia, the Amazon jungle, the Greenland ice sheet, and large parts of western Antarctica will be lost.

If we go over the critical limit of two degrees' rise in temperature, we will have finally lost control of our destiny. The temperature will rise by another two degrees, and when it reaches four degrees, human civilization will collapse. Humanity will not become extinct but will be reduced to a stray herd of desperate survivors. There will not be eight, ten, or twelve billion people on earth, but less than a billion left.

It will not happen in a thousand years. It will happen to those who are young adults today and their children. We will become the ancestors and progenitors of a generation of people who, if they remember us, will curse us as they fight against an overpowering nature that has transformed from a servant into a ruthless executioner.

We have so little time left. It is within the next ten years that our common destiny on earth will be decided.

What is wrong with our way of thinking/reasoning that has brought us so close to the abyss?

As long as there is a life there is hope, and hope nurtures immense damage and destruction.

Knowing how countless people nurture hopeless hopes, illusions, and delusions regarding their lives and reality, it becomes clear why they deny their mental shortcomings and repeated failures and their responsibility to the common good of sustainable future generations.

Rabbi Joseph Telushkin recounts this story of Talmudic logic in his book *Jewish Humor*, illustrating one most crucial mental pitfall affecting many, if not most, people: ignorance combined with irrational thinking.

> A young man in his mid-twenties knocks on the door of the noted scholar Rabbi Shwartz. "My name is Sean Goldstein," he says. "I've come to you because I wish to study Talmud."
>
> "Do you know Aramaic?" the rabbi asks.
>
> "No," replies the young man.
>
> "Hebrew?" asks the rabbi.
>
> "No," replies the young man again.
>
> "Have you studied Torah?" asks the rabbi, growing a bit irritated.
>
> "No, Rabbi. But don't worry. I graduated Berkeley *summa cum laude* in philosophy, and just finished my doctoral dissertation at Harvard on Socratic logic. So now, I would just like to round out my education with a little study of the Talmud."
>
> "I seriously doubt," the rabbi says, "that you are ready to study Talmud. It is the deepest book of our people. If you wish, however, I am willing to examine you in logic, and if you pass that test I will teach you Talmud."
>
> The young man agrees.
>
> Rabbi Shwartz holds up two fingers. "Two men come down a chimney. One comes out with a clean face, the other comes out with a dirty face. Which one washes his face?"
>
> The young man stares at the rabbi. "Is that the test in logic?"
>
> The rabbi nods.
>
> "The one with the dirty face washes his face," he answers wearily.

"Wrong. The one with the clean face washes his face. Examine the simple logic. The one with the dirty face looks at the one with the clean face and thinks his face is clean. The one with the clean face looks at the one with the dirty face and thinks his face is dirty. So the one with the clean face washes his face."

"Very clever," Goldstein says. "Give me another test."

The rabbi again holds up two fingers. "Two men come down a chimney. One comes out with a clean face, the other comes out with a dirty face. Which one washes his face?"

"We have already established that. The one with the clean face washes his face."

"Wrong. Each one washes his face. Examine the simple logic. The one with the dirty face looks at the one with the clean face and thinks his face is clean. The one with the clean face looks at the one with the dirty face and thinks his face is dirty. So the one with the clean face washes his face. When the one with the dirty face sees the one with the clean face wash his face, he also washes his face. So each one washes his face."

"I didn't think of that," says Goldstein. "It's shocking to me that I could make an error in logic. Test me again."

The rabbi holds up two fingers. "Two men come down a chimney. One comes out with a clean face, the other comes out with a dirty face. Which one washes his face?"

"Each one washes his face."

"Wrong. Neither one washes his face. Examine the simple logic. The one with the dirty face looks at the one with the clean face and thinks his face is clean. The one with the clean face looks at the one with the dirty face and thinks his face is dirty. But when the one with the clean face sees the one with the dirty face doesn't wash his face, he also doesn't wash his face. So neither one washes his face."

Goldstein is desperate. "I am qualified to study Talmud. Please give me one more test."

He groans, though, when the rabbi lifts two fingers. "Two men come down a chimney. One comes out with a clean face, the other comes out with a dirty face. Which one washes his face?"

"Neither one washes his face."

"Wrong. Do you now see, Sean, why Socratic logic is an insufficient basis for studying Talmud? Tell me, how is it possible for two men to come down the same chimney and for one to come out with a clean face and the other with a dirty face? Don't you see? The whole question is *narishkeit*, foolishness, and if you spend your whole life trying to answer foolish questions, all your answers will be foolish too."

Now, if you substitute Talmudic study in this joke with "learning to think as wisely and farsightedly as necessary for our long-term survival," you may understand my point!

This is how we often fall in to the trap of human folly, trying foolishly to answer foolish or absurd questions. Human folly is immense and becomes fatal for us in the Anthropocene Age, yet it is invisible for many, if not most, of us, as it is ingrained in our minds as unaware mental programming. And if it seldom surfaces so we can observe it, we prevent any sound feedback on it via cognitive dissonance. One of humans' irons rule is: I am right, you/they are wrong; ergo, I am superior to you.

The next illustrative story on the rabbi and the truth illustrates another blind spot in the realm of human foolishness that creates constant strife, conflict, and division among us.

There were once two merchants who came up to speak due to financial arrears. Both were furious at each other and roc -solid that they each had the right and the truth on their side. Fortunately, they knew a wise rabbi whom both trusted. They went to this one and started talking in each other's mouths. The rabbi suggested that he talk to them one at a time. When he had finished listening to the first one, he said, "You are absolutely right in your arguments, goodbye to you!" Whereupon he sent him home very satisfied.

The other merchant was extremely pleased when the rabbi also gave him the right and sent him home.

The rabbi's assistant, who had heard this interesting exchange of words, went to the rabbi and asked in astonishment. "How can it be that you first gave one right, and then gave the other right?"

"Well, you're absolutely right in what you say too!" replied the rabbi.

Being confirmed partly right doesn't resolve the basic human conflict. The struggle to be right and stamp the other wrong goes on. Much of our folly derives from this mental program. This mental program makes us also dangerously stupid regarding our future prospects. When such mental programs demonstrated in the two jokes can overshadow our self-reflection/critical sense and reality checking, they indirectly repress our capacity to grow beyond our (often petty) differences and act as one common humanity for the common good of all its members.

"All passes away
only change may stay
which can lead to a safe bay
or can lead us astray."

Human folly is expressed in our unsaturated greed, self-indulgence, thirst for power, shortsightedness, and self-deception; ergo, it is a part of human nature, and in our time it has become a real threat for the coming generations' future prospects.

Cheating due to self-interest in the affairs of man is so massive as to render ineffective any regulations/efforts on global climate change.

We lack the capacity for superb thinking. Superb thinking is based on great knowledge of human nature and its pitfalls, extensive knowledge of history, evolution, political reality, and nations' obsession with self-interest. It is contextual, complementing, combinatory, nonconformist, sustainable, farsighted, balanced, and proactive. It leads to advanced, evolving wisdom. Without it we are prone to come up with wishful ideas on how to resolve global climate crisis and other predicaments.

Lack of superb thinking—a rarity among sapiens—implies lack of capacity to focus, to concentrate, to distinguish between essential on the one hand and trivial/petty on the other hand, and being able to execute collective plans beyond immediate personal/national gratification is a huge hindrance to beat human folly.

Are humans genetically inclined to be materialistic to the point of our own destruction?

Is the accumulation of anthropogenic matter merely a measure of humans' annihilation rate? Or will we equip humans with better brains to cope with this problem? These are highly unsettled questions.

Although there is evidence that materialism is learned and shaped by culture, there are some who argue that natural selection may have predisposed our species with a desire to accumulate stuff. Our belongings can offer us a sense of security and status that doubtless played a more important role earlier in human history.

Somehow, creating new stuff has become a possessive word in the collective human psyche. Humans have been conditioned to believe that creating something new is a meaningful purpose of life, and is a very significant way to advance their ambitions.

The limits of science have never been more glaringly apparent when trying to solve this conundrum. Reliance on green technological solutions alone is flawed because the focus is still based on new stuff and more use, not on altering the lifestyles or business models that handed us this problem in the first place. Even if we can replace all fossil-fuel based vehicles with electric ones, for example, cities are already struggling to take road space from cars, and electric vehicles have their own footprint on the world's resources due to the materials needed to build them.

Big technology companies claim they are going green or have set goals for carbon neutrality, but they rarely encourage people to spend less time on social media or order fewer products. Rather, advertising and marketing models convey powerful messages that reinforce the motto "Create and consume more."

This irrational savage materialism is ingrained so deeply with traditions and cultural symbols as well. In the United States, Thanksgiving is followed by another festival called Black Friday. During this ritual, long lines of customers hit the malls and often get injured or trampled, or buy frantically on the Internet. Yet people are convinced that it's an effort worth the trouble.

In the age of Anthropocene, humans may feel entitled to pin hope on technology to fix any problems so that they can continue to do what they are doing. Faced with the accumulation of long-lived plastic in the environment, for example, a spurt of innovation led to biodegradable coffee cups, bags for life, and reusable straws. But while it is true that a sustainable growth model that includes our environment has much larger potential to persist, we need a different approach to sustainability that addresses our massive consumerism.

And in this context comes up another mental program: Denying facts

The passive approach to the proliferation of anthropogenic mass is not merely due to the lack of knowledge about its impact, but in general, it has also to do with human inclination to dismiss facts that don't fit into their worldview. Humans are naturally disposed to disregard issues that are not challenging their daily lives or those that dilute their convenience.

Evidence is accumulating that human-produced hormones are disrupting human and animal reproductive and immune systems. This is affecting the biological development of many species, notably through transgenerational effects or consequences. Estrogen-like compounds can find their way into the reproductive tracts of fetuses and adults, and may affect the sexual identity of people as well.

Many chemicals, both natural and man-made, may mimic or interfere with the body's hormones, known as the endocrine system. Called endocrine disruptors, these chemicals are linked with developmental, reproductive, brain, immune, and other problems. Endocrine disrupting chemicals cause adverse effects in animals.

Studies have shown that, for most species, evolutionary adaptation is not expected to be sufficiently rapid to buffer the effects of environmental changes being wrought by human activity. And our own species will be no exception to this.

While there is no decisive proof that we will destroy ourselves, there are clear indications that we ignore the effects at our own peril. For example, some of the mass extinctions in Earth's history were related to acidification of oceans. The oceans absorb about 30 percent of the carbon dioxide released into the atmosphere, which in turn increases ocean acidity. The oceans may be acidifying faster today than they did in the last 300 million years, primarily due to human activities.

Our impact on the planet is much deeper than carbon footprints or global warming. It points to a future where the effects of anthropogenic matter will take over—if it hasn't already—the identity of Earth and its life. In the face of this, humans themselves might lose out in the evolutionary race.

If you think that people in peaceful countries are protected against human self-destructive folly, just watch a country like Denmark, where I live. The Danish people, instead of waging wars against other nations (well, they did this too to a certain extent through NATO expeditions), live the good life, becoming obese, with lifestyle ailments, being dumbed down, and becoming mentally ill as never before.

My country, Denmark, is considered one of the best and securest places to live in on earth. Social security, medical care, democracy, endless (often futile) debates, and publicly financed institutes have created this country's positive/progressive image. Yet many Danes living in the cities suffer mentally/physically of constant noise hell due to very heavy car traffic. Dramatically growing numbers of them are obese, overweight, mentally fragile, sick, addicts, and homeless. The Danish social model is one of the best, yet 16 percent of the children under age nine suffer from serious mental problems.

Its citizens want to save the world but are not willing to sacrifice their lifestyle/privileges. Every citizen in this paradise produces an average 842 kilograms of refuse. By March 26, 2021, Danish society had already exceeded the quota of refuse produced for the whole year.

Carl Sagan stated once, "One of the saddest lessons of history is this: If we've bamboozled long enough, we tend to reject any evidence of the bamboozle. We are no longer interested in finding out the truth. The bamboozle has captured us. It is simply too painful to acknowledge, even to ourselves, that we have been taken. Once you give a charlatan power over you, you almost never get it back (from *The Demon-Haunted World*).

These are the odds, but to turn around the dice in our favor, we must know who we really are. In chapter 2 we will deal with our expressions in many life areas, illuminating who we really are.

Summary:

In the absence of a folly-proof shield, can we depend on our intelligence in order to survive/thrive after creating such huge damage on the planet? The answer is no way.

Chapter 2

Areas in which human nature/ condition manifests itself with troubling impact

"Humans' fundamental discrepancies: Many of us can talk the talk, few of us can walk the walk!" (B. K.).

"Wise help makes people/humanity balanced, sustainable and evolving. Stupid help does the opposite. It produces parasites" (B. K.).

The human world is partly dirty and corrupted, and the business sector acts often in semi-mafia style. So how are we supposed to resolve our global tribulations with such bedfellows?

We won't, because we can't. Doom on the horizon looms.

—B. K.

Every morning

Every morning
I pray to see my beloved door
behind which what is hidden what I adore,
it makes my soul rise up and soar
leading it to megalopsychia's shore
with meaning of WHAT FOR.
But every morning awaits me
another door
behind which I hear screech and roar,

the hoarse voices of humans' whores.
I wish so hard this door to ignore
but a voice in me insists: EXPLORE!
unless you refuse to be idealistic bore
not perceiving what megalopsychia
has in store.
You have to open this door
to see what behind it: human gore,
and when this filth you will abhor,
will you go in the megalopsychia door...

"These are indisputable facts: 1) the value of human life is not fixed: it is variable/contextual, and it varies from no value at all to high value by contributing to our long-term good and sustainable evolvement; and 2) we don't know whether God really exists or if he is the product of our twisted minds" (B. K.).

The areas in which we often fail

Human mental frailties:
Collective stupidity is the most dangerous threat for us in our present time and in the near future. It is not always malicious as such, but it often results in immense evil and even in irreparable damage.

In the battle to gain constructive self-knowledge/self-determination and our further evolvement, our brains programs/habits can become more and more our Trojan horses. History has taught us an enduring lesson that popular, seductive ideas regarding our future will be found useless without toil, protracted efforts, sufferings and sacrifices.

1. Our minds are split into different "voices"

It is obvious that our most formidable yet invisible foe is our split up and shortsighted sapiens brain, which makes us blind to its fatal shortcomings. A "united being," where all functions manifest themselves through a conscious process, is a rare achievement among human beings.

Cognitive dissonance is a psychological term describing the uncomfortable tension that may result from having two conflicting thoughts at the same time, or from engaging in behavior that conflicts with one's beliefs, or from experiencing apparently conflicting phenomena.

A Jew survived a sinking ship. He swims to an island and
waits for somebody to find him. Years go by and eventually a

ship comes by. He shows his rescuers how he has survived for all these years. They notice that the man erected two synagogues, and ask him if there are more survivors on the island.

"I am the only one," he retorts.

"So why are there two synagogues?" they ask.

"This one," he points toward one of them, "is my synagogue. The other one there, I will never ever set my feet in it..."

Another similar story is about the scorpion and the frog is a crystal-clear example of the split human mind, having the best intentions while sabotaging them either by self-damaging actions or by being good-hearted, naive/stupid, ignoring real danger/dangerous humans.

A scorpion wanted to go to the other side of the river, and therefore went to the bank. Since it couldn't swim, it scouted around for someone who could carry it over the water. Then it noticed a frog that came hopping along. It crawled to her and said, "Hello, my honored and precious frog! I have a small request to ask of you. Would you carry me on your back to the other side of the river?" The frog answered, "I will do no such thing because I know you. I am not a fool, and I know a few things. Don't you think I know that the minute I put you on my back you will sting me?" The scorpion hissed, "Well, the way you reason is anything but sensible. If I sting you, I will drown myself. As you can see, your suspicions defy any reason." The frog protested and maintained that the scorpion was infamous for its unreliability and its choleric temper, but the scorpion kept insisting that it was hardly so crazy as to act contrary to its own interest. Finally, the frog was convinced and let the scorpion climb on it, and henceforth sprang into the water and started swimming. When it had swum approximately halfway, the scorpion stung it in the back. Terminally wounded, the frog asked, just before sinking under water, "Why did you do this? You just told me that it defied any reason!" The scorpion, also about to drown, hissed infuriatingly, "This is not about reason but about character!" And then he disappeared under the water.

Unfortunately, these stories hold a fundamental truth about sapiens as a species either in conflict with themselves or becoming "frogs," transporting sapiens "scorpions" on their backs to some illusionary common destination, or being, again, split scorpions, destroying people in a suicidal manner.

13

Yes, many people are caught up often in contradictions/conflicts, both regarding their statements and actions! The planet isn't going anywhere. Folks with such mental mess. If we don't erase these two or more conflicting mindsets, we're not going to make it as a global, sustainable civilization. And when we're gone we won't leave much of a trace. Another failed mutation; another closed-end biological mistake.

Right now, we are seated on the historic fence and watching carefully where this tide may lead us: to our extinction or to the takeover of an upgraded species, creators much more coherent, wiser, and farsighted than we are. I don't have a clue on how this drama will end up, but I do believe that by leaving the global scene as sapiens, leaving behind our noble essence and letting our upgraded progenies—the creators—take over, sapiens will do the only right thing for the long-term future of ever-evolving advanced, intelligent beings.

2. Addictions/compulsions/obsessions and context-free convictions:

Many people drift aimlessly through life and their chariots have no charioteer to guide them. Human compulsions/addictions could not be documented statistically before our modern time, but it probably existed all throughout our history. In our time, human compulsions/addictions are overwhelmingly widespread. The world has become an ongoing series of new addictive temptations: porn, video games, fast food, social media, (online) casinos, Tinder, designer drugs, consumer products, engineered super cannabis, countless varieties of booze, Netflix, virtual reality, strip clubs, smartphones, cigarettes, the Internet, ubiquitous screens, cryptocurrency, constant new info streams, and so on and so forth.

Most of these things did not exist for the vast majority of human history, especially not in their current enticing forms. It's a minefield out there and only getting more seductive and all-consuming. We're getting so good at creating addictive, attention-hijacking diversions that they gradually dumb people down, making it impossible for them to avoid addictions to various forms of entertainment.

There are over 1.1 billion people smoking, and many hundreds of millions of people are abusing alcohol, drugs of all sorts, medications, food, sex, and engage in regular violent and self-harming/mutilating behavior. In the USA, which suffered in 2018 from a substance (opioid) epidemic, around one hundred people died daily of abuse at the time it peaked. How opioid drugs work explains why humans are so prone to addiction. They suppress pain with impressive efficiency, but they also prime neurons to seek and stay in that state of pain-free bliss. It buries the feelings of pain, anxiousness, and depression so deep so one cannot get upset or alarmed. It seems that it targets both pain and lack of pleasure, which seems to be connected. Their interconnectedness seems to result in addiction (abusing drugs).

Australian researchers have released the first-ever report on worldwide addiction statistics. They found out that approximately 240 million people around the world are dependent on alcohol and about 15 million people use injection drugs such as heroin (1).

Obsessive-compulsive disorder (OCD) affects 2–3% of the population. Obsessive-compulsive disorder is a mental health condition characterized by obsessive thoughts, compulsive behavior, or a combination of both. If on top of it if you put all the people who follow rigid convictions in this world, you may realize that the majority of people are afflicted by one of these conditions.

Obsessions, compulsions, and addictions are the visible indications of troubled minds in humans/societies/civilizations. Countless grownups in modern cities are fixated on being only responsible for their own lives, their comfort and pleasures, without the basic emotional intelligence to solve arising problems and willingness/endurance to shoulder any responsibility for others or the future generations, thus manifesting both emotional/social fixation and infantilization. Many of us are obsessed by the fleeing moment, by religious/social/national/tribal and private rituals, trends, compulsions, and addictions, and have become in this sense semi-automated beings devoid of both critical thinking and overview regarding what is our bearing meaning with our life is (further evolvement to better ourselves/our world).

Excessive self-deception and pretension, which is common for many people, leads to free/critical-thinking repression. Overdoing things is also a widespread form of addiction, like trying to live a healthy life weighing what one eats and adhering to only certain foods regardless of the mental/physical costs. Bulimia, anorexia, registering all food intake are growing compulsory addictions in the world. In my country, lots of people have become obsessed with physical exercise and the right intake of food and calories, causing themselves physical and mental damage.

Many people drift aimlessly throughout life being only steered by their obsessions, be it in the form of impulses or extreme control. The chariot has no charioteer to guide it. Convictions combined with compulsions breed collective foolishness and kill farsightedness—clearly demonstrated in our suicidal era.

Over consumption/pollution/production and population are the most visible evidence of our poor judgement. Instead of acting purposefully and resolutely regarding these dangers, we become prey to our laziness and doubts—a kind of wasteful, utterly useless merry-go-round mental paralysis.

Human addiction is not limited only to the abovementioned afflictions/manifestations. It is overwhelmingly present—the compulsive/obsessive behavior reveals them—in our species' pursuit of gods, greed, wars beside the semi-automated, repetitive, replicating rituals, actions, and convictions. It is present in our struggle for status, power, and riches; in our groupthink; in our craving for endless gossip, scoops, sensations; in being immersed in banality, triviality, daydreams, and "sweet nothing" wastes of time on account of dealing with vital/menacing issues. People are very often unaware of their addictions as their self-deception mechanism delude them to believe that they do possess free will.

If a person could live 200 years/was equipped with better brains than the average sapiens, he would consider most of what we utter as superfluous, distracting, and futile.

The best deal one can strive for in a lifetime is to have proactive ideals, mission/vision to work for and pursue, and reality that challenges you to work together with others for higher goals than your own petty life and engage yourself in it as to better it by unfolding your best potentials. How many of us do just this?

Human folly due to context-free thinking:

The following story illustrates our inclination to think context free.

A man caught a bird in his snare. The bird told the man that he, who had eaten so much beef and lamb in his lifetime, couldn't satiate his appetite by eating such a tiny bird. The bird added that if the man released him, he would give the man three pieces of wise advice. The first one would be given while the bird is standing on his hand, the second while standing on the roof of the nearby house, and the third while standing on a branch of the nearby tree. The man consented, and the bird, standing on his hand, said, "First advice: do not believe in absurdity no matter who tells you." The bird flew up and landed on top of the roof. "Second advice: do not grieve over what is past and finished. It's over. Never regret what has happened." And then the bird added that inside his tiny body was a huge pearl weighing ten gold coins, and that the man, by releasing him, had lost this fortune with which he could have fed himself and his family for the rest of their lives. The man began wailing violently. The bird said, "Didn't I just tell you not to grieve for what is past and over? And didn't I tell you not to believe in absurdity? My entire body doesn't weigh as much as ten gold coins. How could I have a pearl that big and heavy inside me?" The man realized that the bird was right and stopped crying. "Fine, I can see it. Now please give me the last piece of advice." "The last piece of advice is this: don't give advice to somebody who is groggy and falling asleep. Don't throw seeds on the

sand. Some torn places cannot be patched." (Idris Shah, *Tales
of the Dervishes*).

If people could think in shifting contexts mode, would they defy such
stupidity, which is a kind of subtle tyranny destroying our lives on earth? In the
world of shifting contexts, destiny is not always fixed but something that we and
our circumstances are actively involved in without compulsory control our rituals.

One of the most / intractable and compulsory problems of men, is believing
too much in -isms. Religion is not the only one to offer people context-free
conviction: capitalism, liberalism, humanism, socialism, communism, and
other -isms do the same work on human minds: they instill in them rock-hard
convictions that reduce free thinking. The price tag to pay if you are a very devout
believer in some religion or ideology is that you must give up our grandest gifts:
the capacity and right to think critically, independently, rationally, and inquiringly
beyond the framework of groupthink; the right and capacity to defy unfounded
truths; and the right to be free willed. If your faith has brought you to the harbor
of fundamentalism, you may have become a danger to yourself and to others. You
risk becoming an *agelast*, a person who does not hear God's laughter. God laughs
at anybody, according to a Jewish saying, if they take themselves too solemnly and
practice their truths as the only truth on earth, to the point of exerting atrocities
against infidels. God laughs bitterly at such persons because they do not just
miss the point of what our ultimate meaning in our grand journey is all about,
but because they come to harm and damage much more than they ever can heal
and build up.

The following Jewish story may illustrate my point.

Jossel and Messiah

Many years ago, there was a little remote village that was
inhabited by Orthodox Jews. Because the village was far away
from the main road and from the self-imposed tumult of man,
the inhabitants enjoyed a quiet and peaceful life without big or
nerve-wracking events. Even so, they had their little worries,
being human as they were, and one of these worries had grown
rather big throughout the years. Since their village was situated
far from the beaten track, people were concerned that their
Messiah would overlook them when one day he comes to release
the Jews from their suffering and agony. This thought made
them anxious, but being practical people, they searched for an
appropriate solution. They built a watchtower on a small hilltop
on the outskirts of town and hired Jossel, the fumble finger of
the village, to be on the lookout for the passing Messiah. If he

should pass and not notice the village, Jossel was to make sure that he did and would invite him to stay. One day, when Jossel was dozing off in his watchtower, his old friend, the town loafer, came to visit him. "Yes, yes." The friend nodded appreciatively. "It is a job of great responsibility you've got there, my friend. You must be very pleased indeed!" Jossel pondered it for a second while rubbing his eyes. "Pleased? Yes, I guess I'm rather pleased! I might not be making a lot of money doing this, but at any rate, it's a safe profession."

Jossel and people like him serve a significant purpose for believers by sitting on top of their watchtowers and watching out for some savior for us. They may elevate the believers' spirits and morals, insisting on pacts between God and us. As long as they do not become a source of strife and hatred between people—which they often become—they are blessed, because they bring hope and strength to those poor souls who wish to believe in some deep meaning to their miserable lives. But this is not always the case with religiosity. Often people interpret their religion—as our era shows very strongly—as a reason to look down on others who do not share the same convictions and thus they have the holy right to pursue and persecute them.

3. We tend to view ourselves as omnipotent. False, bloated self-view creates vainglorious self-harming situations.

Our brains are designed in such a manner that we often overestimate ourselves and our brains, thus stepping on our own feet. False, bloated self-identity creates imaginary realities. It is expressed by the swagger attitude. The definition of swagger is an overly confident and arrogant walk or mannerism. The haughty, confident walk of someone who knows he is good-looking is an example of a swagger.

The faith in our omnipotence has other expressions as well:

It is said that the king of Babylon built a maze that was so ingeniously arranged that even the wisest men did not venture into it, and those who did too quickly went astray. One day the king of the Arabs came to his court, and to make the guest's simplicity a laugh, the king of Babylon made him go into the labyrinth. The Arab king wandered around confused until darkness fell. Then he asked for God's help and found the gate. When he came out again, he told the king of Babylon that if he wanted, he would one day show him an even better maze. He traveled back to Arabia, gathered his troops, defeated Babylon, and captured the king. Then he bound him on a fast camel and led him out into the wilderness. After riding for three days, he said to the king of Babylon, "At home you would let me perish in a bronze maze full of stairs, doors, and walls. Now the Almighty has wanted me to show you mine, where there are no stairs, doors, corridors, or walls blocking your way." Then he untied the king's ribbon and left him in the middle of the desert, where he died of hunger and thirst (Jorge Luis Borges).

And the moral? Humans know a lot but nature/the universe determines their fate and can therefore not be ignored or destroyed without it involving man's own downfall.

Aesop told the story of the frog and the ox apropos omnipotence.

"Father," said a little frog, "I have seen such a terrible monster! It was as big as a mountain, with horns on its head, and a long tail, and it had hoofs divided in two." "Tush, child," said the old frog, "that was only Farmer White's ox. It isn't so big either. He may be a little bit taller than I, but I could easily make myself quite as broad, just you see." So he blew himself out and blew himself out and blew himself out. "Was he as big as that?" asked he. "Oh, much bigger than that," said the young frog. Again the old one blew himself out and asked the young one if the ox was as big as that. "Bigger, Father, bigger," was the reply. So the frog took a deep breath and blew and blew and blew, and swelled and swelled and swelled. And said, "I'm sure the ox is not as big as that." At this moment he burst.

The human illusion of ineffability, of never being wrong. It is often expressed in the three dictums of religions/ideologies:

1) We are right.
2) You are wrong.
3) We are therefore better than you.

If you know people well, including mass media influencers and politicians, you realize how often they fall into this trap, as well as their admirers. And when they are exposed for misjudging or failing badly, they are judged harshly by their former followers and become bitter, angry, depressed, and often die shortly afterward. (In Israel's modern history it happened to Moshe Dayan, to Israel's chief of staff David Eliezer during the Yom Kippur War, and also others.)

The "I have the right on my side and you are wrong, ergo I am better than you" trap makes us stupid by viewing reality as black/white, right/wrong. Maxim Gorky wrote a story on the outcasts: senseless drunkards, thieves, loafers, cheaters. And yet when they started to discuss local subjects and politics, they turned their vodka-poisoned minds into the right ones, the fresh future spirit, while the shopkeepers and the merchants were not real humans. So stupid can people become that even when they have lost anything, including their dignity, and destroy themselves, they may still think that they are better than others and they are going to change the world with their drunken speech. People need not be drunk in order to make themselves feel worth more than others by loose talk.

The west is expert at conveying this stupid attitude by preaching its morals to the rest of the world, all the while its societies and people are in paralyzing crisis. Twenty years of bloody, fruitless, and futile war in Afghanistan; the devastating, costly, and monstrous wars in Iraq, Libya, and partly in Syria waged by the west (NATO) have brought about mass death and immense destruction, not a single enduring positive development. All the while our societies are plagued by exploding obesity, lifestyle ailments, mental problems, poverty, and dumbing down of people. As conspicuous consumers, we have not cut down on neither our lavish lifestyle nor our profane moral superiority. We keep being, in this century, the so-called authorized moral preachers to all the others, while contributing to tension and strife around the world. The west moralizes China, Russia, and other countries in a shameless style, and they return with the same coin. Russia is an emerging power (2022) and it can't be intimidated as China can't be intimidated, so why waste time instead of finding the critical area—environment and detente—where we can cooperate with them? Because we can't. We can create divisions, tensions, and conflict by working globally, but unifying for a great goal we can't. This is the brutal expression of human nature: power games based on division, and the powerful dominate. There are no good ones and bad ones in this game of futility, just dirty human reality and nature...

There is nothing impossible for a man who is not forced to follow words with actions.

Oh, blinded sapiens!
In Paradise on earth you believe
while yourself and others you deceive
—B. K.

The propensity for human omnipotence can be viewed as the Dunning–Kruger effect, which is a hypothetical cognitive bias stating that people with low ability at a task overestimate their ability. As described by social psychologists David Dunning and Justin Kruger, the bias results from an internal illusion in people of low ability and from an external misperception in people of high ability

Extreme views often stem from people who feel they understand complex topics better than they do. Now, as ever, societies need to know how to combat this kind of folly.

Folly is using a rule where adding more data doesn't improve your chances of getting a problem right. In fact, it makes it more likely you'll get it wrong. Intelligence combined with experience, resulting in wisdom, on the other hand, is using a rule that allows you to solve complex problems with simple, elegant solutions. Folly is a very interesting class of phenomena in human history, and it has to do with rule systems that have made it harder for us to arrive at the long-term life-affirming solutions. If I were asked what's the greatest problem facing the world today, I would say folly. Folly is the core reason for averse global climate change, arms races, military conflicts, poverty and many other problems bothering us.

The psychology of people being unbending optimists in the face of adversity is incredible in its naked blindness. Most people are not aware of this core reason and they blame the wrong causes. Often, it takes a near apocalyptic disaster to wake up the sleeping majority. In everyday life, the same punishment is required when a person denies his own limits. People, generally speaking, will blame everything but their own unawareness and deficient self-knowledge.

4. We are prejudiced and judgmental and become easily mental slaves due to massive social/religious/ideological brainwashing.

Humans are highly social creatures. Our brains have evolved to allow us to survive and thrive in complex social environments. Accordingly, the behaviors and emotions that help us navigate our social sphere are entrenched in networks of neurons within our brains. Social motivations, such as the desire to be a member of a group or to compete with others, are among the most basic human drives. In fact, our brains are able to assess "in-group" (us) and "out-group" (them) membership within a fraction of a second. This ability, once necessary for our survival, has largely become a detriment to society.

Understanding the neural network controlling these impulses, and those that temper them, may shed light on how to resolve social injustices that plague our world. People are prejudiced, sometimes unashamedly so. We tend to have a host of reasons ready to justify our biases: the mentally ill are dangerous, immigrants steal jobs, the LGBTQ community corrupts family values, Muslims are terrorists, and rural whites are uneducated.

From cultural, visionary twentieth-century forums we ended up in an era of shitstorms orchestrated by raging flocks of human swarms. Why so? One of humans' tragedies is that they need to accentuate differences with others and create divisions in order to define themselves.

You may remember the ugly elections in the USA in 2020. The choice was between Biden and Trump, and the two camps used stinking shitstorms against each other, employing many disguised and dirty tricks in fighting each other. In Israel, something similar happened in the elections in 2021. Democracy seems to hang on the ropes regarding decency and fairness as divisions are being accentuated to the point of prejudice. Being so prejudiced and condemning is a part of human foolish inheritance. "Them"—the wrong ones—and "us," with the patented truth. And shows itself both in small as well as huge groups. It poisons many relations, including international ones, often sabotaging any possible comprehensive cooperation between nations.

The politicians we choose must project confidence, overview, and power, and often find an enemy in order to accentuate their messages, even when they lack everything. Why do people keep electing such politicians? Confident people are more convincing. This has been demonstrated in many studies. Politicians are clearly aware of it, hence all the media training and PR management. Any politician that doesn't come across as assured and confident gets (metaphorically) destroyed. So confidence is important in politics, even though often being fake.

However, the Dunning-Kruger effect reveals that less-intelligent people are usually incredibly confident. More intelligent people, by contrast, aren't at all. Self-appraisal is a useful metacognitive skill but one that requires intelligence. If you don't have much of it, you don't consider yourself flawed or ignorant, because technically you don't have the ability to do so.

Studies have also shown that when a confident person is shown to be wrong/ lying, they are then considered far less reliable or trustworthy than an unconfident person. This may explain the negative image of politics, which is mostly a series of (often pretending to be) confident individuals making big promises and failing miserably to keep them. That sort of thing really puts people off.

Confidence can be a useful facade, and confident appearances matter. However, high intelligence can be a hindrance in some areas, making a person hard to understand in some settings or overly analytical. And knowing one's own ignorance contributes to this feeling!

5. We are under the spell of *mundus vult decipi.*

"I mean that part of the game of beautifying our reality follows the same principles of the emperor's new garments. We wish subconsciously to hear/follow/pursue soothing and self-glorifying fairy tales on our existence/ourselves" (B. K.).

"There is this huge vanity market selling illusions for people by flattering them, and lots of human parasites live on it" (B. K.).

"Two miracles occurred with a 2,000-years gap: Jesus had turned water into wine. We turned countless mediocre consumers into unique beings" (B. K.).

Miracles and lying are interconnected twins. Lying is a definite human propensity. It is so common in the human world because so many people will buy lies blindly. This is what *mundus vult decipi* is all about.

A very substantial mental program in our mindset is ***mundus vult decipi, ergo decipiaur***(The world wishes to be cheated and therefore it has been cheated), and our self-view and notion of our competence are badly based on it and thus contaminate our overview and strategic thinking.

A current delusion is that Wall Street will save us from our global climate crisis by investing heavily in green energy. Is it possible that a burglar turns out to be our guardian, or is it again one of our mind spins? A way to get away from self-delusions and baseless hopes is to look closely on both human nature and on the facts of our real reality.

Besides our irresistible hang to pursue *mundus vult decipi, ergo decipiaur* which involves also shortsighted thinking due to our propensity to be wishfully cheated on a huge scale, it is the fuel of the law stating that anything that can go wrong in the human world will go wrong. This is mainly due to our incapacity to think farsightedly and life-affirmingly over generations.

An illustrative example for *mundus vult decipi* is our attempt to resolve the menacing self-created climate problem. It is self-evident to the wise that unless we reduce our consumption patterns, including self- indulging goods, and our numbers on the planet, thus consuming much less energy and goods, this problem will get out of control. Nothing less will do the trick.

The second example is the assumption that consuming less polluting energy by technological improvements will make energy cheaper. Since 1960 the use of fuel for a passenger per kilometer on a passenger plane went down by 75 percent. It meant cheaper flights, and voila, the collective emissions of CO_2 by planes exploded. We have become more effective in using energy since 1960 but the emissions of CO_2 in this period has grown threefold.

The third example is the phenomena resembling pricking a balloon. By moving heavy industry from the west to Asia and mainly China, we brought down our coal use by 8 gigawatts, while China used more than 43 gigawatts and it is developing coal-powered facilities with the capacity of 121 gigawatts. That is to say that China produces goodies with much more CO_2 emission than the west produces.

The fourth example is storage of CO_2 under earth or the oceans. We have invested in developing carbon catching and storing facilities around the world. USA emits 1.2 billion tons of CO_2 annually. So far man has only stored 25 million tons annually. After many years of developing this mechanism, we store only a minuscule portion of the world's CO_2 emissions.

The fifth example: J. P. Morgan presented a calculating piece in which it is stated that every year 5 billion cubic meters of oil is pumped up, transported, and refined. If we wish to capture and store around 6 billion CO_2 annually, which is around 15 percent of the global emissions, we are forced to build up such a huge infrastructure in order to catch and store this CO_2 volume, and which will be much bigger than all the production and transportation and consumption of the whole oil industry. Building up such an infrastructure behemoth will result in extreme huge CO_2 emissions.

The sixth example of *mundus vult decipi* is the business of electrification, which goes very well in Norway (water energy and financing the project through oil funds) and in France (nuclear energy), but the rest of the world is far behind. And only 18 percent of world energy production is by electricity, while the sale of electric cars globally was 2.5 percent of all sold cars in 2019.

The seventh example is that huge oil and gas companies invest more and more in green energy as they are financially encouraged to do so. This means much more produced green energy, lower life standards for us all (as energy will be much more expensive than other fossil sources), and the process will go very slowly. IAE claims that even by 2040 70 percent of our energy will come from manufacturing / production. The biggest consumer of fossil sources is industry, which uses huge amounts of gas. If it turns to electricity, the energy expenses will grow by three- to sixfold without the investment expenses. To this fact we have to add that a big chunk of the production industry cannot use electricity, either because it demands high temperatures or the raw materials used are full of carbon [steel, cement, ammoniac (as in fertilizer), and plastic]. All need fossil sources or very high temperatures. The production of these four is expected to rise up.

A pressing question that many ignore is whether green energy is really green. The production of green energy will demand smooth and non-polluting mining industry, but this branch is far behind regarding this energy shift. Research shows that the mining of materials needed for electric car batteries must grow by 87,000 percent, raw material to wind energy by 1,000 percent, and to sun cells by 3,000 percent from 2016 to 2060. This will affect our environment badly, with much poisonous garbage (from producing solar panels alone, the pollutants will grow from 250,000 tons in 2016 to 78 million tons in 2050).

Can we trust these figures? As the report was written by Michael Cembalest, a top figure in strategic investment in J. P. Morgan who manages 2.2 trillion dollars, it is hard to believe that he would allow wrong figures in this report. Now the question is whether we can escape the collapse of our civilization due to worsening climate and lack of global sustainability without cutting down rather drastically on our consumption, pollution, and population. I don't think it can be done!(2)

6. People have a patent on the right faith/ideology.

People who think their opinions are superior to others are most prone to overestimating their relevant knowledge and ignoring chances to learn more. Researchers distinguish "belief superiority" from "belief confidence" (thinking your opinion is correct). Belief superiority is relative. It is when you think your opinion is more correct than other people's. The top end of their belief superiority scale is to indicate that your belief is "Totally correct (mine is the only correct view)."

Across five studies, Hall and Raimi* found that those people with the highest belief superiority also tended to have the largest gap between their perceived and actual knowledge. The belief superior consistently suffered from the illusion that they were better informed than they were. As you might expect, those with the lowest belief superiority tended to underestimate how much they knew.(3)

An example from life conflicting with the faith of humanism and multiculturalism:

1. The failed integration of many Muslims in Europe is due to stupid assumptions that people from entirely other culture/religions can be integrated into their host countries with different cultures, values, and dominating religion.

And the outcome of such assumption/faith: twenty retired French generals have called for a military takeover in France in 2021 if President Macron fails to stop the country from "disintegrating with Islamists." The open letter, which was signed by more than 1,000 officials in the military, police, and gendarmerie, was instigated by retired gendarmerie officer Jean-Pierre Fabre-Bernadac. First published on his blog on April 14, it was republished a week later by right-wing magazine *Valeurs Actuelles*.

"France is disintegrating with the Islamists in hordes in the *banlieues* (suburbs), detaching large parts of the nation and turning them into territory subject to dogmas contrary to our constitution," it adds. The outcome of this multiethnic/religious experiment is growing tension, friction, and probable future bloody conflicts.

Faith in something beyond one's own ordinary life is essential for most people, especially because it fortifies and solidifies one's notion of identity and worth. But faith, be it in God or ideology, without 1) deep knowledge of the human shifting and complex reality and 2) deep knowledge of human nature and our propensity for division and mental limitations to build a bridge over them is like having two blind people (you and your faith, the other and his faith) trying to find the way together and both falling into the muddy pool. Most human folly gets its nourishment from this blindfolded combination.

7. Eddie's human dilemma: talking but not acting. The idiocy of theorizing without possessing the capacity to "swim."

Eddie is my good friend living in Israel with his family. He is as old as I am (at the moment, when I write these lines, I am seventy-six years old), and we grew up together from the age of sixteen in Kibbutz Negba in the southern part of Israel. Ever since, even when I left Israel in 1972 for good, we've kept in touch. And every time I visit Israel, we meet. Eddie is burdened by a dilemma. He wants to make a difference in his world but does not act upon this wish in a way that gratifies him.

His dilemma is very common and widespread among sapiens: the wish to act upon an idea or plan and then failing to do so. Why is Eddie's dilemma so common? It is so because my friend Eddie's dilemma is a symbol of the nightmarish dilemma of the modern Job. Like Job in the Bible, Eddie has gotten all the material prerequisites to enjoy a good, extraordinary life. Yet contrary to the story of Job, Satan does not destroy his lifestyle and hit his family. What blocks Eddie's way is his wavering free will and lack of focus and determination. He identifies the necessities of his time and knows that we, as a civilization, pass through a very perilous stage in our journey. He knows that it is an extraordinary era because we risk destroying our life conditions on earth (if global temperatures keep on rising). He knows that this era demands extraordinary people to rise to the challenge. He understands the urgency, yet he is entangled in his own local/regional struggles, which goes nowhere and dissipates his energy, making him passive for not making a difference where we are exposed to mortal danger. He asks all the right questions: Is acquiring material gratification all that life is about? Is it the purpose of our lives to grow old, play golf or bridge, and become nice grandparents and simultaneously destroy our grandchildren's future?

His dilemma is how to give his life meaning and purpose beyond his own private/local/regional spheres of interest. Eddie, like a billion people in the modern world, has been granted the greatest material/educational/vocational privilege in human history: learning to think independently and to identify the great impending necessities of our time and to be able to choose freely how to use his free time and energy. But he does not do the right thing.

Many people have a hard time lifting their heads up to see other horizons than the ones they are already acquainted with or have been conditioned to viewing. This constitutes compulsive behavior, which dangerously ignores the shifting challenges in our ever-changing reality. We see it clearly in our private lives. We often would ditch our promises to ourselves and others in favor of our own short-ranged interests and short-term economic pragmatism, security, and convenience, regardless of the long-term consequences of this strategy.

8. The negative mental legacy.

There are so many disturbed and immature people who become parents and who damage their offspring psychologically for life. What percentage of kids have bad parents? A study found that 20 percent of moms and dads have been called "bad parents" (September 17, 2019). What makes a parent unsuitable? A parent may be deemed unfit if they have been abusive, neglectful, or failed to provide proper care for the child. A parent with a mental disturbance or addiction to drugs or alcohol may also be found to be an unfit parent. Being extremely poor as to have difficulties sustaining a family, resulting in the stunted growth of children is also a huge problem.

Unsuitable parent results in early child emotional deprivation, which damages people for life and often reduces their chances of becoming emotionally balanced and developing appropriate conflict solving traits.

In 2020–2021 a poisonous debate rattled Danish society. The kernel of it was whether the country should forbid circumcision of babies and boys (in reality, Jewish and Muslim babies) up to the age of eighteen, when they can legally make their own decisions. The argument was that circumcision is an act of changing the body of the child without his/her consent. There was not much consideration for the religious implications of such a draft, which could easily turn such an act into part of civilizations colliding. Neither was consideration given to what extent we harm many more babies and children in the west by letting them grow up with unsuitable parents and harm many hundreds of millions of people's bodies, health, and lives with our unfair resources sharing, where hundreds of millions people don't get enough food or our excessive/damaging food industry and consumerism that turn many of us into obese, sick people. But allowing unsuitable parents to get children is the main reason for how we pass on the negative social legacy from one generation to the next, harming both more than 20 percent of all children and our societies as well.

9. The ways we view things as existing or nonexisting.

Does God exist if the Holocaust could take place? When people prove to be so satanic and merciless, how can God exist and not prevent them from doing their inhuman misdeeds? Most religious people reconcile themselves with the fact that life can be terrible and God does not always protect them as he should (according to their faiths) for his own invisible reasons. But not having God at all, regardless of his seemingly ambivalent behavior, is worse than having amputated God.

What should not exist does not exist. And what should exist, does it exist?

Escapism is bound to reject real reality: An anecdote illustrating this has been told about the sultan of Constantinople who decided to conquer the island of Malta because it was supposed to have a great strategic significance. He sent off his admiral and a mighty flotilla to take Malta by force. The admiral sought and sought after Malta but could not find the island. He turned the ships around and sailed back to Constantinople. As he was called on by the sultan to inform him of the expedition, the sultan inquired eagerly whether he had succeeded in capturing Malta. "No, my honorable ruler!" the admiral answered. "It is because Malta is *yok*!" (*Yok* means "nonexistent.")

When people feel that their "Malta" is *yok*, they cannot come up with adequate solutions to their lives' problems. They become helpless and powerless. It is rather petrifying to experience how often people's mode of thinking works toward denial of reality, which often elicits an escapist response in us. In both religious and ideological faiths, we see human convictions overpowering our reason and observations, making us "beautifying" shamelessly the unseen "out there," even though it cannot be proved or verified (which reminds me of the story "The Emperor's New Clothes" by H. C. Andersen). Religious or ideological and hedonistic forms of escapism are shared by most people on earth, and have generated delusion-based norms as well as unsustainable humanity.

When a distorted norm backed up by the majority of people does not match with reality, the believers, in order to avoid inner conflict and conflict with reality, simply extend reality, claiming that there is much more in life that we don't perceive by observation. Like the religious disciples explaining the drive for careless multiplication of families by the argument that the dining table of God will take care of their needs. It is semi-smart but shortsighted escapism! Denial of reality, escapism, contributes immensely to making people ignorant and stupid. People don't enjoy being ignorant, but they are not aware that they are. Denial of real reality and escapism are, therefore, tightly bound to each other as psychological mechanisms, and they foil any attempt to attain clear sight, oversight, farsightedness, and skillful strategic conflict or problem-solving in the mind of their host.

10. Debasing our fine ideas by becoming mentally fossilized, compulsory or vain.

When God created the earth, he was delighted by and pleased with his masterpiece. He looked at the beautiful earth, the green grass and trees, the colorful flowers, the roaming animals, the blue sky, and the white clouds, and was proud of his feat. Then the devil stealthily arrived and stopped beside the Almighty. The devil clapped his hands with great enthusiasm and exclaimed, "This is incredible, fantastic, and splendid. Only you, the Almighty, could create such a thing." He paused, looked at the Almighty, and whispered, "Wouldn't it be a brilliant idea to institutionalize this?"

The cunning devil in us takes our godlike, beautiful creations and turns them into something debasing, context free, stiff, burdensome, and boring by introducing endless repetitions, compulsion and rituals, greed and power monger's lust. The schism between our drive to create something new on the one hand and our inclination to preserve our achievements and constructs unchanged also undermine and destroy them and sometimes also us.

Nasiruddin, ferrying a pedant across rough water, said something ungrammatical to him. "Have you never studied grammar?" asked the scholar. "No!" "Then half of your life has been wasted." A few minutes later, Nasiruddin turned to the passenger. "Have you ever learned how to swim?" "No. Why?" "Then all your life is wasted. We are sinking."

Vanity is excessive pride in or self-admiration in one's own appearance or achievements. It makes people think and act as if they were much better, stronger, and wiser than they and others are, and their convictions are unassailable. Human history is littered with much evidence of this overestimation of self/one's convictions, and therefore it keeps damaging our fine efforts in this world.

I once read a little parable about a man who spent much of his life in a prison cell. One day, he was paroled and sent out of prison to face a world he did not know, and it frightened him with its vastness and lack of order. He rented a little room, which resembled his prison cell, and spent most of his time in this room, going out only in order to shop. One day, on his way back from the marketplace, he found an empty bottle and a wire, which he took with him to his room. Safe in his room, he pressed the wire into the bottle. After some days, he took the bottle up and crushed it. The wire landed on the ground, crooked and twisted.

The point of this allegory is that once you enslave people's minds for a long time (often to the point where they become uncritical, ritualized, overconfident, and comfort seeking), they find safety in this form of enslavement and will often keep reproducing it through their behavior and thinking. Many people who are not capable of self-reflection that brings self- knowledge reduce their life to mediocre existence. The problem facing them is while they try to flee from difficult situations and controversial decisions, their tensions keep building up because they need much energy to avoid and ignore what they don't dare face. They are like the child who ran away from his home. He ran around his home and was on the verge of fainting when a passerby stopped him and asked him what was he doing. The child told him that he ran away from home and his family. The stranger asked him, "So why do you run around the house time and again?" To this question, the child replied, "I am not allowed to go over the road out there."

11. Becoming bored by life, thus wasting our precious time on pleasure, triviality, and petty things.

Boredom in many humans' lives is a kind of compulsion that forces them to find an escape gate in the form of pleasure, triviality preoccupation, and vain self-focus. This malaise, making countless people waste their time on escaping boredom instead of engaging proactively and life-affirmingly in life, is described by the following story.

> The Almighty and Peter walked in Nablus and saw a woman sitting in front of her house, being bored to death. She was so profoundly bored that the Almighty felt pity for her, and out of his pocket he drew up a hundred lice, spread them on her, and said, "Rejoice, my daughter. Now you have something to occupy your mind with!" And so it went true: killing the lice made her glad, and fighting them, mad.

Much of the human curse is due to the itch not to get bored, as most humans don't know how to keep themselves engaged and ablaze in a life-affirming, long-term manner without the crutches of consuming lust, virtual games, or some sect or ritual self-stimulation, which often dumb down their minds. They need distractions, and their lack of courage to make an essential proactive change in their routine-prone life ends up in compulsive behavior.

12. Modern diet, overweight/obesity and lifestyle ailments.

Worldwide obesity has nearly tripled since 1975. In 2016, more than 1.9 billion adults eighteen years and older were overweight. Of these, over 650 million were obese. Thirty-nine percent of adults aged eighteen years and over were overweight in 2016, and 13 percent were obese (June 9, 2021).

Obesity and getting dumber

Being significantly overweight comes with plenty of well-documented health problems already. From the risk of injury due to stressed joints to type-2 diabetes, obesity isn't just dangerous, it's deadly. Now, a new study reveals that obesity might actually also affect your brain, and not just in a "I should grab another doughnut" kind of way.

In a new research paper published in the *Journal of Neuroscience*, researchers reveal that lab testing in mice has hinted at diminished brain power in animals who are dramatically overweight. The fat and not-so-bright rodents may actually be incapable of processing information as well as their more fit counterparts.

As *Science News* reports, the researchers used two groups of mice for the study. The first group was plumped up with a high-calorie diet for three months, while the other group was kept at a standard calorie intake. The obese mice, which weighed nearly 40 percent more than the control group, showed diminished reasoning and memory skills when confronted with the same puzzles as the control animals. The larger rodents had problems remembering how to navigate mazes that were trivial for the mice that had remained at a normal weight. To determine what might be causing the mice to become dimmer, the researchers began looking at their brains. What they found was that the obese mice had a greater number of microglia, which are a type of "scavenger" cell that serve the immune system in the central nervous system. These cells aren't "bad," per se, but an increase in their numbers is thought to cause damage to the tiny stalks on nerve cells that receive messages. These stalks, called dendritic spines, are incredibly vital for the brain to process information, and less active spines may actually be inhibiting the ability of the obese mice to process information as rapidly as their healthier kin.

It's obviously not clear at this point whether the same is true in humans, but the functions of the brain that are of concern in this study are fairly similar between rodents and many other animals, including humans.

Modern diets make us sick

In 2006, Prof. Aryeh Levin, currently director of the Pediatric Gastroenterology Department at Wolfson Hospital and a leading researcher in inflammatory bowel disease, asked himself exactly the same questions. "Something in the accepted theory that Crohn's is an autoimmune disease in which the body attacks itself does not make sense to me," he says in a special interview with *Mint*, "because this theory does not explain why the disease is rising in western countries, why in the Third World it's typical of the population in high socio-economic status. My suspicion then was that the cause of the disease was food. The food we eat today produces epidemics that the body is unable to deal with."

Indeed, the genetic theory, which was blatantly illogical in Prof. Levin's opinion, was refuted in 2012. At that time, the most prestigious scientific journal, *Nature*, published a genetic study whose results stunned scientists. Seventy-five thousand subjects, also from Israel, participated in the study. It was found that less than 25 percent of the population carry the problematic genes for the development of inflammatory bowel disease. But the really interesting finding was that people who do not have inflammatory bowel disease (the control group of healthy people) also carry these genes. This study showed that only a minority of people develop these diseases on the basis of a genetic background.

13. The modern excessive self-reflection leading to human neuroticism.

It can be said with impunity that never before our time have there been so many self-reflective people who try to understand their motives, desires, and mind and to try reign them in so as to become mentally balanced, physically healthy, and wise. Yet this huge number of people don't seem to have any impact on bettering our world or making us change our unsustainable ways of living/consuming. Furthermore, many of these self- reflective people are plagued by mental disturbances and problems. Often, the medicine seems to turn on them. How come?

Listed below are some citations made by Heraclitus. All four of them bear deep truths that most of us have forgotten to follow in our time:

"Much learning does not teach understanding."

"Abundance of knowledge does not teach men to be wise."

"Nothing endures but change."

"Whoever cannot seek the unforeseen sees nothing for the known way is an impasse" (from *Fragments*).

Heraclitus' observations are very relevant to our time, especially the lack of understanding, wisdom, and a new invigorating global vision, and can explain why we are in such great trouble globally. We have all the information we need on the state of the world, but being what we are, we cannot transform this information into operational wisdom. We have too many people living on this planet consuming and polluting much too much, thereby tipping the planet's sustainability and life-bearing conditions. We have too much economic growth/activities that threaten our global sustainability and life prospects in the future, and we lack a unifying global guiding vision to save us from our misdeeds.

Expedient "transactional" amorality is the loudest voice of our time. It means taking the quick, easy solution that is in one's self-interest and operates without underlying, long-term bearing principles. This means that people do reflect over their deeds, but lacking the wisdom to do so, they reach views and decisions that are disastrous for future generations. We hear *transactional* frequently in the news in the USA, being used as a trendy word for lack of long-range planning or vision that can save our progeny from global catastrophes. These are undeniable facts, and we don't focus on why we often act against our long-term interests when we reflect! Therefore, our future is likely to be very bloody, with immense suffering and mass death of both humans and animals before we learn to reflect sensibly and realize these reflections on a long-term basis. In this process, we've, alas, passed the opportunity to find elegant global solutions, a la win-win, to avoid the oncoming self-inflicted torments.

14. The humans' blind adoring mentality.

Three pious Jews—one from Poland, the second from Holland, and the third from America—met and started bragging about the merits of their respective rabbi. The Polish Jew tells of his rabbi, "One day a devouring fire broke out and threatened to burn down the whole city. And do you know what our rabbi did? He went into the middle of the street, where the fire was ablaze, and roared, 'Fire to the right, fire to the left.' And by doing it, he saved the city."

"That is a very great miracle indeed," said the Dutch Jew, "but listen to the great miracle my rabbi performed. One day the sea broke through the dikes. Great waves of water gushed through and threatened to drown the whole country. And do you know what my rabbi did? He went to the scene, stood firmly in front of the huge waves of water, and roared, 'Water to the right, water to the left.' And by doing it, he saved the country."

"Amazing," said the American Jew. "You have very powerful rabbi, but listen to the amazing miracle my rabbi achieved. He went on a plane from New York to Chicago. And just before the Sabbath commenced, he got a telephone call from some influential connection telling him he could earn much money if he concluded a deal right now in New York. What could the poor rabbi do? It was now the holy Sabbath, where a Jew is not allowed to travel or work. Then he stood up and roared, 'Sabbath to the right, Sabbath to the left.'" (A joke I heard told by the rabbi of Copenhagen.)

Piously, blindly admiring people are to be found by the legion. In the American elections in January 2021, the Americans were divided into two camps, both more or less blind disciples of the two candidates who fought each other with such ferocity and hatred that the idea of civil war was mentioned. Both candidates exposed very severe mental and moral flaws that only blind disciples—and they were more than 150 million of them who voted for them—could believe in their statesmanship. Blind adoration followed by hatred and harassment of the other party is the utmost expression of human stupidity, and it shows up in elections, racial hatred, religious divisions, persecution of minorities, nations, and regarding men and/or women.

Blind admiration/loyalty in the human domain is thus an expression of human stupidity and should therefore be addressed as the core cause for both division and hatred among us.

15. The overblown myth of free will.

The focus/faith of our modern times on individualism, our free will, and self-realization paradoxically often brings about mental misery and ennui. Critical thinking, self-knowledge/knowledge of the human state/the world are preconditions for exercising farsighted free will, and only farsighted free will aiming at maximizing our progenies' survival and evolution is real free will. Impulses, routine, and daily conformism don't point clearly in this direction.

Have you ever spent a bunch of time going down social media rabbit holes or reading the latest outrage-inducing "news" stories only to feel that you wished you'd spent that time differently? Media is one of the best examples of an industry whose integrity has been destroyed by the incentives built into capitalism. In order to be profitable, social media and news sites require huge masses of people to look at the advertisements on their websites. As a result, the top priority of these companies becomes maximizing 1) the number of eyeballs on their websites at any given time and 2) the number of hours every pair of eyeballs spends looking at their websites.

Unfortunately, to maximize traffic, news companies have learned that it's best to be polarizing, controversial, emotionally intense, and sensational. Clickbait titles that distort the truth are used to trigger our limbic system, our primal anger/ fear response, causing us to click, read agitatedly, and get embroiled in flame wars in the comments. The result of this alliance, for vast herds of unsavvy users, is living life in a near-constant state of dissatisfaction and agitation, itching to pick up the smartphone to find out how the "idiotic socialists" or "right fascists" are ruining the country today. Much of this drama and outrage is simply manufactured.

"Splitting hairs" or "one cannot see the forest for the trees"(lack of overview) often used by the mass media and social media has reduced the manifestation of clear thinking and free will among people, creating both status granted and indulgent consumer mentality and unbearable and confusing inferno, which disguise this mental slavery from the consuming masses.

Strict, rigid faith grants both security and stupidity but not so much free will, while free/critical thinking generates doubt but also access to some free will if used wisely. The conforming gravitational forces of society, its groupthink, cut off whatever free will wings they try to nurture for most people.

Petty, envious, hyper-consuming people won't be able to focus on megalopsychia, which implies the exertion of free will to form a promising future.

Taking orders is easier for many than being self-governing.

Following context-free convictions may make many people accepting of orders and reducing their free will, as they are disguised compulsions, closing people's minds off to sound/needed feedback, making them thereby act semi-automatically. Experiments show that 90 percent of what people do in a day follows routines so strictly that our behavior can be predicted with just a few mathematical equations. These findings are part of a psychological work that shows human behavior in a weirdly mechanical light. Some psychologists claim that most of a person's everyday life is determined not by his conscious intentions and deliberate choices, but by mental processes outside of his awareness and triggered by stimuli in his environment. In other words, most of the time we simply react instinctively and through fixed mental patterns to the world around us. It appears that although we think we are reasoning out our decisions and choosing our actions deliberately, we may often just be responding more or less automatically to cues in our environment. Only afterward do we make up reasons to explain what we did. Our public space is filled with repetitive, obsessive chatter and chipper from people writing and singing popular music, of commentators interpreting the fleeting moments, of politicians and mass media repeating as talkative parrots their worn-out slogans. The quibbling people of the mass media world, dissect pettiness-prone subjects with their mini needle sword. The biggest contributor to the infantilization of the masses, thereby reducing whatever free will they possess, is the mass media/TV coverage and social media through their constant focus on sensationalism and frantic focus shifting.

But claiming that we lack free will altogether is wrong. I wrote the following lines long time ago, breaking the rules of human determinism by exerting great degree of free will.

"The ultimate human Meaning and Must,
is in defying the destiny of animated dust.
Challenging the constraints of life's Animator,
to evolve and ascend, and become a creator"

Yet free will leading to long-term wise and life-affirming decisions and actions is limited commodity in many human affairs, including in dealing with our global predicaments. The experience of possessing free will, the feeling that we are the authors of our choices, is so utterly basic to everyone's existence that it can be hard to get enough mental distance to see what's going on. On the deepest level, if people really understood what's going on, it's just too frightening and difficult unless we do something radical to enhance this frail free will. Otherwise, for anyone who's morally and emotionally deep, it's really depressing and destructive. It would really threaten our sense of self, our sense of personal value. The truth can be just too awful.

To sum it up, we must aspire to cultivating our frail free will by transcending the limitations, rituals, and compulsive/conviction prone human mindsets of sapiens. We have not come that far. If we follow this course and save ourselves from our own blind, self-destructive actions, we will prove that we can enact our free will. But it is not truly operational in our current times

16. The idea that human nature is benign.

If you expect love, compassion, and harmony to reign one day in the human world, you are delusional about what sapiens are.

Many, if not most, people tend to idealize the premises of our existence under a compassionate God's supervision or some benign universal energy flow, while reality points toward both more murky human nature and nature itself. They view themselves as God's children with aspirations for love, compassion, and harmony. But is there such thing as God? And if he exists, is he compassionate at all? Are we compassionate under all circumstances? Billions of *Homo sapiens* who have sought refuge in the vague idea of us being good, loving under the supervision of some Almighty God have caused irreparable damage throughout human history, yet they seem to relish the misconception that nature is benign and we are benign, loving, compassionate beings. Facing the cruelty of life, they claim that we have gone astray by following a sinful way.

The cruel fact is that nature is indifferent, neutral, almost cynical—let the fittest survive. Violence is also in our nature as it is part of the survival repertoire of all animals. Sometimes you have to kill in order to eat; defend yourself and your short-term interests; possess others' land, property, and remove competition and survive. We are not noble, kind, tender, and loving savages who got spoiled by our culture. We have been dangerous killers who have contributed to the extinction of many people, nations, species, though maybe not the Neanderthal, elves, and hobbits. We are inherently as violent and vicious as we are loving, kind, and peaceful. Nature has given us the capacity and potential for both.

The fittest surviving in a Darwinian sense is a matter of species resonance, not pure violence for its own sake. Yet violence in itself prevails and is all dominating.

17. Modern societies create vulnerable, infantile personalities while mental suffering increases.

The West is devoid of a great Quest!
The Greek historian Polyb wrote the following: That no civilization has

surrendered itself to an aggressor if it had not previously developed the ills that ate it up from within.

Mental health conditions are increasing worldwide, and especially in the west. The reasons can be because of demographic changes, aging populations, and lifestyles that predispose youngsters to becoming mentally sick. There has been a 13 percent rise in mental health conditions and substance-use disorders in the world in the last decade (up to 2017). Mental health conditions now cause one in five years lived with disability(4)

Why are mental health problems increasing? Rates of mood disorders and suicide-related outcomes have increased significantly among adolescents and young adults, and the influence of modern life, digital communication, mass media, and social media are all to blame. Mental health problems are on the rise among adolescents and young adults, and individualism/narcissism, digital dating accompanied by "meat market," and reduced capacity to create deep relations, all supported by the mass media/social media, may be drivers behind the increase.

The working market also plays a role in increase fleeting and unstable lifestyles. A new study found that 32 percent of Gen Z respondents say they are the hardest working generation ever, and 36 percent believe they "had it the hardest" when entering the working world compared to all other generations before it.

Does infantilization of children and teenagers also play a major role in the growing mental malaise?

The dictionary defines infantilizing as treating someone "as a child or in a way that denies their maturity in age or experience." What's considered age-appropriate or mature is obviously quite relative. But most societies and cultures will deem behaviors appropriate for some stages of life but not others. As the Bible puts it in 1 Corinthians 13:11, "When I was a child, I talked like a child, I thought like a child, I reasoned like a child. When I became a man, I put childish ways behind me."

Some psychologists will be quick to note that not everyone puts their "childish ways" behind them. You can become fixated at a particular stage of development and fail to reach an age-appropriate level of maturity. When facing unmanageable stress or trauma, you can even regress to a previous stage of development. A child is taken care of but learns slowly to take responsibility both in regards to himself and the people around him. The child learns that reciprocal relations are built upon being both mutual and responsible. As the modern phenomenon of growing numbers of young people without any responsibilities to others than themselves spreads like a fire, so has the infantile personality of many of them become accentuated.

Many societies nowadays produce infantile people in the backdrop of crumbling families and extreme individualism.

Can entire societies succumb to infantilization?

The biggest contributor to the infantilization of the masses is the mass media/ TV coverage that shifts focus constantly. Newspapers/mass media focus on hair splitting actuality on account of strategic thinking regarding our direction as an evolving species. Frankfurt School scholars such as Herbert Marcuse, Erich Fromm, and other critical theorists suggested that, like individuals, a society can also suffer from arrested development. In their view, adults' failure to reach emotional, social, or cognitive maturity is not due to individual shortcomings. Rather, it is socially engineered.

If we counted all the different infantilized addicts in a modern society based on creating/enhancing needs like sex, shopping, pornography, food, drugs, cigarettes, Internet games, and booze, they would be a majority, telling us clearly what kind of monstrous civilization we are supporting.

Lots of people don't have clue on how to mature, often due to a combination of bad mental capacity, shitty emotional backgrounds that conditioned them to be blind to their faults and mental problems, making them self-focused, rigid minded, and even worse, compulsive-obsessive.

Modern societies create obsessive consumers' mentality

"People in democracies being dumbed down by commercialism, consumerism, trivia and trifles, follow this dumbed-down course, being steered by massive commercial brainwashing tyranny" (B. K.).

What is modern consumerism?

Consumerism is a concept that is shaped by social and economic conditions. Consumerism by itself is part of the general process of social control and cultural hegemony in modern society. Addicts of modern consumerist culture erode the humans' spiritual connection with nature and with each other as social mutually dependent beings.

How does consumerism affect our society?

As well as obvious social and economic problems, consumerism is destroying our environment. As the demand for goods increases, the need to produce these goods also increases. This leads to more pollutant emissions, increased land use and deforestation, destruction of the seas, and accelerated climate change.

There were 11,973 people (up to April 1, 2021, a year after being launched) who saw the program *Shopping for Freedom*, documentary on the history of advertising and consumerism on YouTube. The program describes a terrible process used by companies in order to promote selling of their products. It is partly done by sending subliminal messages to the unaware people, thus making them inclined to buy certain products and turning them into obsessive/compulsive consumers—a mental condition most people in the west suffer from. On the other hand, influencers of buying lifestyle and social prestige by connecting them with certain goodies are scoring millions of watchers.

What percent of people in US have a shopping addiction?

A 2016 meta-analysis suggests that about 4.9 percent of Americans are addicted to shopping. The prevalence is even higher among university students (8.3 percent) and among shoppers (16.2 percent).(5)

How many people in the US have compulsive buying disorder?

More than one in twenty adults nationwide suffer from compulsive buying, according to a telephone survey of 2, 500 adults(6)

Subliminal stimuli to manipulate emotions?

Research suggests that it is possible that the subliminal stimulus that you can't see can influence your judgment of the stimuli that you can see. Subliminal messages in advertising are designed to engage people subconsciously. These ads use various colors, shapes, and words that enable customers to make small but powerful associations between a brand and an intended meaning. What these advertisers do is manipulate people to desire unnecessary goodies and status symbols as real needs. They have created a desiring culture/mentality, turning the manipulated desires of billions of people into needs. The same manipulation and creating desires/dreams takes place in the political life to a great extent, while the development of critical thinking in the population is neglected, making many people into manipulated beings.

Doesn't manipulating people to be compulsory consumers or following seductive political ideas mean dumbing them down and making them semi-automatic beings? No doubt about this too.

The chicken get hypnotized!

Those of us who have acquired advanced knowledge in mass psychology including mass brainwashing and manipulation, know that these mechanisms—which we, humans, are very susceptible to due to the fact that we are social animals—are used all around the world, including in our so-called enlightened democracies.

They should glue and coerce individuals together around a common narrative and consensus. Religion does the trick by repetition of prayers, by replication of so-called truths. Our democratic institutions, of course, massively use this mechanism of repetition and replication in our everyday life in order to fortify our convictions that we are the best, as democracy is the best political form of the future, and we are so advanced as to choose our leaders, discard them, and hence have become the wisest political beings. As with prayers, it may take longer than a lifetime to get evidence to support these claims, but in the meantime... the chicken get hypnotized!

18. The overblown, overpopulated, encroaching, dense, and mobile humanity threaten our future and dumbs us down.

We count by now 8 billion people on earth (2022), and many of them are concentrated in mega cities and other huge cities, impacting their environments and themselves in a rather alarming manner. For the average person, a day of twenty-first-century urban life basically consists of moving through a concrete, mechanized maze of flashing neon signs, massive billboards, whooshing automobiles, police sirens, construction noises, car horns, and hundreds of disinterested smartphone-gazing people—more people than our hunter-gatherer ancestors would've seen in their entire lives.

The twenty-first century is characterized by a tsunami wave of supernormal stimuli, and the modern metropolis is the epicenter. In these places there is often a pervasive, vague, eerie sense of falseness, artificiality. Living relatively isolated lives in manmade environments, we become disconnected from each other, from community, and from the natural world. Disconnected from ourselves, each other, and nature, we (unconsciously) search for anything to numb ourselves or feel a fleeting high. And as we've already seen, supernormal vices are eagerly waiting to ensnare us.

For much of human history, in diverse cultures worldwide, life was viewed as sacred. Nature, along with all the gifts it provided, was sacred. Life moved much more slowly and was a much quieter affair, allowing people to be deeply in touch with the sounds, seasons, rhythms, and therapeutic beauty of nature's processes of growth and decay. People lived close to the land, and the aliveness of nature and everything in it was an ever-present, immersive reality. There was magic in nature, in the mysterious forces that animated the whole nature.

"God is dead," Nietzsche was quoted for, referring to the death of God in the hearts of western people, to the disenchantment of the world. This is what it was like to be human for much of our history. When you compare this vision of life with that of the modern age, it's easy to see how far we've drifted from our roots. Humanity is with its back to the abyss, fighting its self-created immense global problems.

And so comes the environmental fallout

The world accumulates heat now (2021) twice as quickly than in 2004(7)

Huge iceberg in Antarctica is getting loose (2021(8):

All these averse changes due to our growing population, consumption, production, and pollution.

What was the Earth's population in 1950?

In 1950 there were 2.5 billion people on the planet.

What is the world's population in the beginning of 2022?

The current world population is over 8 billion. The term "world population" refers to the human population (the total number of humans currently living) of the world.

In a matter of seventy-one years the global population tripled. Our energy per capita doubled, so we now consume at least six times more energy than in 1950. Population growth is associated with the premise of economic growth, which destroys life conditions on Earth. So far only Japan has followed cautiously the course of dwindling population without any catastrophes to its life standard.

What was the approximate annual average CO_2 level in 1950?

Up until 1950 the levels of atmospheric CO_2 were pretty steady at 300–310 ppm (that's parts per million out of all the molecules in the air). In 2021 it stood at 404 ppm. It is also clear that our consumption of the planet's resources and pollution/ destruction of the natural habitat have increased dramatically in these seventy-one years as to tip our global climate/temperatures. Take for example a country like Iran, which reflects the dire state of many countries (https://m.ynet.co.il/articles/ HylEgaDo00). The Islamic state in the Persian Gulf is facing an existential crisis due to overexploitation of groundwater due to growing population/consumption of food and other commodities, and a continuing decline in precipitation

The sources of fresh water in our warming world are running out. In many countries, especially those that were not blessed with rain in the first place, the rise in global temperatures translate into a decrease in the annual amount of precipitation and a reduction in the water reserves available to the population. At the same time, global warming and the continuing increase in the human population are increasing the demand for water for agriculture, which is further increasing the pressure on dwindling water reservoirs.

Such a severe water crisis is expected to occur in Iran, among other places, according to a study published in the journal *Nature* (2021). The researchers, three Canadian scientists of Iranian descent, examined the changes in groundwater volume in Iran in 2002–2015, based on data released by the Iranian Ministry of Energy. Their findings show that the combination of declining rainfall and overexploitation of groundwater reservoirs for the needs of its growing population is degrading the country toward an environmental catastrophe bordering on real existential danger.

Desert countries such as Iran, where living conditions are so close to the limit of human existence, are particularly sensitive to global warming. In the last hundred years, the average temperature in Iran has risen by about 1.8 degrees, 1.5 times more than the average global rise. The amount of precipitation falling on it per unit area is less than a third of the world average, while the evaporation of water from the surface is three times greater than the average. The rise in temperature is reflected in the accelerated loss of surface water, the worsening of droughts, and the growing shortage of fresh water for drinking and agriculture. The distress is reflected, among other things, in the increase in the number of wells for pumping groundwater in the country. Between 2002 and 2015, their number increased by more than 70 percent, from 460,000 wells to 794,000, while its population grew by only 17 percent.

Despite the significant increase in the number of wells, the total amount of water pumped each year from most of the country's aquifers, its natural groundwater reservoirs, has actually decreased. On first reading this figure may sound encouraging, assuming it indicates an improvement in Iran's water resources management. But the reality is quite the opposite. According to the researchers, this is clear proof of overexploitation of groundwater. The depletion of aquifers is particularly significant in agricultural areas, where more than half of the water used to irrigate the fields is pumped from the depths of the earth.

If the aggressive exploitation of groundwater continues, Iran is expected to face a particularly deep environmental and socioeconomic crisis in the near future. The first danger poses to the very food security of its residents, a fifth of whom are employed in the agriculture industry. The Iranian desire to abolish its dependence on other countries and to agricultural independence, born in part as a result of international sanctions imposed on it, may even have exacerbated the water crisis by encouraging greater use of groundwater for irrigation and food production.

The natural environment is also harmed. Due to the low water density in them, freshwater aquifers usually reside over saltier and denser aquifers. Over-pumping from the sweet upper aquifer reduces the pressure exerted on the saline reservoir deep from it. As a result, the salt water spreads upward and pollutes the soil with salts. Indeed, in some agricultural provinces in Iran where groundwater consumption is particularly high, a significant increase in soil salinity has been measured, which could impair its fertility and also exacerbate the country's food shortage.

And if that is not enough, the disappearance of the aquifers affects the mass balance of the soil, causing it to compact and sink. Land subsidence due to groundwater loss is well felt throughout Iran, including its capital Tehran, whose surrounding areas are sinking at a rate of about 20 centimeters per year. Soil compaction reduces the capacity of groundwater reservoirs, and impairs soil stability in earthquakes and thus also increases the potential for damage to buildings and infrastructure due to seismic activity.

Sustainable management of water resources in the face of climate change will be one of the major challenges facing the world in this century. The challenge is likely to be particularly difficult in developing countries that lack the economic and organizational capacity to deal with it successfully. Many are located in arid regions of the Middle East, North Africa, and Southwest Asia. Along with the need to adapt to global changes, many of them are also experiencing rapid population growth and rising living standards, which will only increase the demand for the dwindling vital resource. (9).

19. The modern world makes us mentally ill and dumb.

Physicians and statisticians who analyze patterns of births in the United States have concluded that the number of babies born with some physical or mental defect has doubled over the last twenty-five years (July 18, 1983).

Are birth defects on the rise?

The latest report from the Centers for Disease Control and Prevention finds that the number of cases of birth defects nearly doubled from 1995 to 2005. The problem has continued to increase since 2005 for babies born to mothers of every race and age group (January 22, 2016).

Are ADHD numbers increasing?

A new study finds ADHD diagnoses in children between the ages of four and seventeen increased from 6.1 percent in 1997–1998 to 10.2 percent in 2015–2016. "This is a dramatic change," explains study researcher Wei Bao, MD, PhD, an assistant professor in the College of Public Health at the University of Iowa (November 26, 2018).

Does ADHD affect a certain population?

The American Psychiatric Association (APA) says that 5 percent of American children have ADHD. But the Centers for Disease Control and Prevention (CDC) puts the number at more than double the APA's number. The CDC says that 11 percent of American children aged four to seventeen have the attention disorder (October 11, 2017).

The prevalence of autism in the United States has risen steadily since researchers first began tracking it in 2000. The rise in the rate has sparked fears of an autism "epidemic." But experts say the bulk of the increase stems from a growing awareness of autism and changes to the condition's diagnostic criteria (March 2, 2017).

The modern world is wonderful in many ways (dentistry is good, cars are reliable, we can so easily keep in touch from Mexico with our grandmother in Scotland), but it's also powerfully and tragically geared to causing a high background level of anxiety and widespread low-level depression. Soren Kierkegaard, a Danish philosopher, claimed that the deeper you suffer from anxiety, the greater you are (probably thinking of himself, glorifying his misery). Now we have more people than ever people suffering from anxiety in Denmark and in the west. I treated many of them, but I must be blind since I could not notice their greatness.

Is mental illness more prevalent in our time?

Mental illnesses are common in the United States. Nearly one in five US adults live with a mental illness, 46.6 million in 2017 (February 1, 2019).

What percentage of the world suffers from mental illness?

Around one in seven people globally (11–18 percent) have one or more mental or substance use disorders. Globally, this means around one billion people in 2017 experienced one. The largest number of people had an anxiety disorder, estimated at around 4 percent of the population (May 16, 2018).

There are seven particular features of modernity that have psychologically disturbing effect. Here they are:

1. Meritocracy

Our societies tell us that everyone is free to make it if they have the talent and energy. The down side of this ostensibly liberating and beautiful idea is that any perceived lack of success is taken to be not, as in the past, an accident or misfortune, but a sure sign of a lack of talent or laziness. If those at the top deserve all their success, then those at the bottom must surely deserve all their failure. A society that thinks of itself as meritocratic turns poverty from a problem into evidence of damnation, and those who have failed from unfortunates into losers.

The cure is a strong, culturally endorsed belief in two big ideas: luck, which says success doesn't just depend on talent and effort; and tragedy, which says good, decent people can fail and deserve compassion, rather than contempt.

2. Individualism

An individualistic society preaches that the individual and their achievements are everything, and that everyone is capable of a special destiny. It is not the community that matters; the group is for no-hopers. To be "ordinary" is regarded as a curse. The result is that the very thing that most of us will end up being, statistically speaking, is associated with freakish failure.

3. Secularism

Secular societies cease to believe in anything that is bigger than or beyond themselves. Religions used to perform the useful service of keeping our petty ways and status battles in perspective. But now there is nothing to awe or relativize humans, whose triumphs and mishaps end up feeling like the be all and end all.

4. Romanticism

The philosophy of Romanticism tells us that each of us has one very special person out there who can make us completely happy. Yet mostly we have to settle for moderately bearable relationships with someone who is very nice in a few ways and pretty difficult in many others. It feels like a disaster in comparison with our original huge hopes.

The cure is to realize that we didn't go wrong; we were just encouraged to believe in a very improbable dream. Instead we should build up our ambitions around friendship and nonsexual love.

5. The Media

The media has immense prestige and a huge place in our lives, but routinely directs our attention to things that scare, worry, panic, and enrage us, while denying us agency or any chance of effective personal action. It typically attends to the least admirable sides of human nature without a balancing exposure to normal good intentions, responsibility, and decency. At its worst, it edges us toward mob justice.

The cure would be news that concentrated on presenting solutions rather than generating outrage, that was attuned to systemic problems rather than gleefully emphasizing scapegoats and emblematic monsters, and that would regularly remind us that the news we most need to focus on comes from our own lives, direct experiences, and our civilization's long-term evolvement and survival.

6. Perfectibility

Modern societies stress that it is within our remit to be profoundly content, sane, and accomplished. As a result, we end up loathing ourselves, feeling weak and sensing we've wasted our lives.

A cure would be a culture that promotes the idea that perfection is not within our grasp, that being mentally slightly (and at points very) unwell is an inescapable part of the human condition. And that what we need above all are good friends with whom we can sit and honestly discuss our real fears and vulnerabilities on top of other perspective/actions beyond our narrow self for the common good of our future generations.

7. Smartphones and social media

Smartphones and social media don't just affect individuals, they affect groups. The smartphone brought about a planetary rewiring of human interaction. As smartphones became common, they transformed peer relationships, family relationships, and the texture of daily life for everyone, even those who don't own a phone or don't have an Instagram account. It's harder to strike up a casual conversation in the cafeteria or after class when everyone is staring down at their phones. It's harder to have a deep conversation when each party is interrupted randomly by buzzing, vibrating "notifications." Life with smartphones means "we are forever elsewhere."

All young mammals play, especially those that live in groups like dogs, chimpanzees, and humans. All such mammals need tens of thousands of social interactions to become socially competent adults.

How does living in a city affect your health?

The detrimental effects of urban living on physical health have long been recognized, including higher rates of cardiovascular diseases, cancer, autoimmune ailments, and respiratory disease. More recent, however, is the revelation that urban living can also have adverse effects on mental health.

How does the city affect people's brains?

The brains of people living in cities operate differently from those in rural areas, according to a brain-scanning study. Previous research has shown that people living in cities have a 21 percent increased risk of anxiety disorders and a 39 percent increased risk of mood disorders (June 22, 2011).

How does the city hurt your brain?

Just being in an urban environment, they have found, impairs our basic mental processes. After spending a few minutes on a crowded city street, the brain is less able to hold things in memory, and suffers from reduced self-control (January 2, 2009).

Cities are associated with higher rates of most mental health problems compared to rural areas: an almost 40 percent higher risk of depression, over 20 percent more anxiety, and double the risk of schizophrenia, in addition to more loneliness, isolation and stress.

What does loneliness do to us?

Chronic loneliness increases the risk of heart disease, high cholesterol, and diabetes. It can cause mental health problems such as anxiety, emotional distress, addictions, or depression. Loneliness can also increase the risk of suicidal death.

Does loneliness cause death?

Loneliness was associated with higher rates of depression, anxiety, and suicide. Loneliness among heart-failure patients was associated with a nearly four times increased risk of death, 68 percent increased risk of hospitalization, and 57 percent increased risk of emergency department visits(10).

Loneliness can be deadly, this according to former surgeon general Vivek Murthy, among others, who has stressed the significant health threat. Loneliness has been estimated to shorten a person's life by fifteen years, equivalent in impact to being obese or smoking fifteen cigarettes per day (March 20, 2019).

Loneliness is harmful for our brains and cognitive capacity.

The Internet surfing/game for the sake of keeping boredom at bay, is a Faustian bargain, whose endgame is loneliness and dumbing folk down! What is meant by a Faustian bargain? A Faustian bargain is a pact whereby a person trades something of supreme moral or spiritual importance, such as personal values or the soul, for some worldly or material benefit, such as knowledge, power, or riches.

There is a dramatic growth in loneliness through frantic social mobility/ avoidance of close relations, plus obsessive reliance on the Internet. There is also a dramatic growth in loneliness in the world through social isolation/avoidance. There are estimated over one million young Japanese who isolate themselves in their homes and the plague is spreading to both middle-aged and older people. They call this self-isolation phenomenon *hikikomori*. It is an extreme reaction against extreme pressure from Japanese society regarding education, work life, and social life, all long, demanding, and ritualized. This trend may become common as it spreads in the world.

As robots and machines will make a huge chunk of human work redundant, IT jobs could be dealt with from people's home and the Internet can hold their growing army of hermits amused/entertained. An the modern phenomenon of loneliness can become both contagious and mentally/cognitively debilitating. Unfortunately there are no gains, only losses, in this self-isolating 'lifestyle. A future where robots and smart machines will make most of our work will probably be a nightmarish life for most people stamped as useless(11).

Who does eco-anxiety affect?

In the more serious cases, eco-anxiety can cause a sensation of suffocation or even depression. Among the latter group, it is quite common for people to express a strong sense of guilt about the situation of the planet, which can be aggravated among those who have children when thinking about their future. Eco-anxiety is exploding. "Eco-anxiety" refers to persistent worries about the future of the Earth and the life it shelters (September 22, 2020).

How many people have eco-anxiety?

An APA poll conducted at the tail end of 2019 revealed that more than half of adults in the US reported having "at least a little 'eco-anxiety,'" and "nearly half of those aged 18–34 (47%) say the stress they feel about climate change affects their daily lives" (June 8, 2020).

20. People become mental slaves due to massive social/religious/ ideological brainwashing.

Many studies in the field of psychology and sociology illustrate public passivity in the face of brutality bordering on mental slavery by referring to an incident that occurred in 1973. In this incident, two robbers entered a bank in Stockholm, Sweden, with guns and dynamite, took four hostages—three women and a man—and held them hostage for 131 hours. After their rescue, the hostages showed a peculiar behavior. These people who had been threatened, abused, and intimidated felt gratitude toward their captors and tried to protect them when expert investigations were made. One of the women became emotionally attached to one of the assailants, and another began a campaign to raise funds for the legal defense of the criminals.

As strange as it sounds, similar situations as the robbery occur in daily life with abused children, battered women in relationships, prisoners of war, victims of incest, and generally for many people being abused/brainwashed in our societies. The explanation lies in our survival instinct, described here as Stockholm syndrome, and our groupthink as well. When the lives of victims depend on the action of their assailants, the emotional reactions of some victims turn into gratitude once they survive, just as slaves may have also expressed gratitude when they were given their freedom. Similarly, in many contemporary families the victims, feeling hopeless, develop positive feelings toward the abuser or controller, and rationalize their acceptance of such behavior while reacting negatively to family or friends who try to rescue them, and have difficulty freeing themselves from this emotional trap.

Groupthink is also a part of this reaction, as detaching yourself from what you consider as the safe, normative, and protecting is very dangerous for most persons.

For mental slavery to occur, there are five typical situations:

- Perception of a threat, physical or psychological, and the conviction that misfortune can really occur
- Appreciation of small acts of kindness by the abuser toward the victim
- Isolation from others
- Conviction that one is unable to escape the situation
- Pressure to conform to the norms of groupthink and not deviate

A shopping advertisement is an example of "soft" enslavement of people's minds used by commercial brainwashing agents. It is a malignant globally allowed practice by our societies, promoting consuming advertisements to brainwash people to consume. In the United States, 36.5 percent of adults are obese. Another 32.5 percent of American adults are overweight. In all, more than two-thirds of adults in the United States are overweight or obese (June 2, 2020). Can it really be that this food orgiastic consumption was caused by people who were not enslaved mentally in some manner?

How many people in the western world converted their desires for the good life—like excessive shopping, sex, travelling, living luxuriously compared with humans living one hundred years ago—into perceived needs, thus becoming mental slaves of our commercial brainwashers? We will never get the right figures of this malignant phenomenon. Simply because our economic machine depends on spending and buying excessively, and admitting that it enslave lots of people, will be censored and denied by the commercial institutions that manipulate the human souls. The figures presented on compulsory shopping will therefore be much lower than in reality, and the system will consider it as a personal problem, as usual, when its premises come under scrutiny.

During Passover of 2021 in Copenhagen, Denmark, one could see long lines of consumers waiting to go into open shops after some of the COVID-19 restrictions were lifted. Those lines tell the real story of the mental slavery lots of people have sunk into.

In the real narrative of the Jews who left Egypt, two forms of slavery were presented and challenged by Moses. The first was the Jews' slavery in Egypt, and the second, Korah's attempt to stage a coup the mission toward the Promised Land with the temptation of the golden calf. Moses resisted both.

The core of excesses leading to mental slavery is inherent in humans, but the stimuli to excesses has to do with a society manipulating this core for its own reasons: to keep the citizens pacified, docile, and falsely content with their patented good lives and ideology. This is a prescription of manipulating people to become consuming slaves.

When we march together toward attaining a demanding common goal or a future, we can keep focused and united as long as this mission is our sole focus. Once it is away and people/nations view their lives through their personal/national perspectives, division and turmoil will ensue, followed by mental/material and sensual decadence. In Israel, building the nation demanded a common front against its enemy. The enemy outside and the ultimate purpose kept the unity and the focus of this young society. When there is no common demanding, great goal and disruptive forces threatening the common project, people easily become slaves to their whims.

21. We exaggerate in all, including tolerance toward the very rich and human rights, which results in abuse and social rerouting.

What kind of human justice are we accepting when Oxfam says the wealth of the richest 1 percent is equal to the other 99 percent (published January 18, 2016).

The richest 1 percent now had as much wealth as the rest of the world combined in 2016, according to Oxfam. It used data from Credit Suisse from October for the report, which urged leaders meeting in Davos to take action on inequality. Oxfam also calculated that the sixty-two richest people in the world had as much wealth as the poorest half of the global population. It criticized the work of lobbyists and the amount of money kept in tax havens. Oxfam predicted that the 1 percent would overtake the rest of the world soon.

What kind of human justice are we practicing in our countries? New Jersey's most populous city paid tribute to George Floyd, unveiling a 700-pound bronze statue in his honor outside Newark's city hall. After six burglaries, three car thefts, committing two violent home invasions, three armed robberies, beating four victims senseless, passing counterfeit money, and being arrested twenty-three times since 1998, George Floyd did not commit a crime in over one year before being killed by the police.

Nancy Pelosi, in June of 2020, gave the Floyd family a folded American flag that had flown over the capitol on the day he died—as a fallen hero (July 18, 2021, Walter Luce, Facebook).

Is it human nature to exaggerate? Embellishment is part of human nature, experts say, and almost everyone is guilty of it at one time or another. Left unchecked, however, exaggerations that seemed innocuous at first could result in serious, potentially bad-ending consequences.

Does exaggerating mean lying? We all have a tendency to exaggerate. It makes our stories funnier or more dramatic. After all, when you exaggerate, you're not really lying, you're just overstating things. The word *exaggerate* can also suggest that a particular characteristic is overdone or almost larger than life.

Exaggerating also means making principled decisions based on a certain context, but being out of touch with changing contexts. In such cases the decisions based on exaggerations do become self-defeating, as the case is with human rights (without counterweight in human duties and commitments to act sustainably and justly), certain people accumulating much of the world buying power or hyper liberalism (not setting any limits for people's consumption, pollution, and numbers)

A friend of mine wrote this little excerpt, expressing vividly how our highly praised tolerance turns many of us into stupid beings by its excesses and exaggeration.

Praise of our western tolerance

Let us all praise Tolerance, because Tolerance will tell us how tolerant we are, while others are not. It tells us that there is no human rights or good things neither in Russia nor in China.

There was never any tolerance there. That's just how it is. Remember it!

You should not believe that Trump was not the worst president of US presidents before or after. If you say such a thing—that he was not worse than many others—your acquaintances will change the subject, and will greet you next time from a distance when you see each other.

You must not sympathize with the sufferings inflicted upon "Me too" men. It is not up for discussion!! Are you a man as you try to focus on their sufferings instead of their misdeeds? You'd better watch out next time you meet a woman!!!

Are you a woman? Remember that if you are a woman you are a strong one!! Always ready to fight! But do not despair as there are seventy-one different genders, which one are you today? Be open minded. Be You. Be different!!!

Remember not to tell your boy that he is a boy. Remember to tell your daughter that she can become a boy if she feels so. The possibilities are many when we are so hellish tolerant.

The ironic thing about the far left and academia, and now corporations, is they have recreated discrimination and divisiveness in the name of diversity, inclusion, and equality. A whole, highly paid industry has grown up with consultants who run "training programs" to try to teach white males to hate themselves and feel guilty for things they never did or thought.

They accompanied this with all the rhetoric about racism, and all the other isms, and indoctrinated the students with the idea that all white males are oppressors, and there is systemic racism. They combined this with the need for the cradle-to-grave welfare ideology, which preaches that wealth created through hard work and brains is bad, and the proceeds of that effort needs to be given to those who have neither. So today we even have the Pentagon stressing inclusiveness and equality over military skills.

PS: If you are in doubt about your opinions expressing tolerance, better keep them to yourself. Or try gently to consult with someone you trust. But beware and comfort yourself with the fact that tolerance is patient, tolerance is mild, it does not damage anything. It does nothing indecent, does not seek its own agenda, does not get angry, does not bear grudges, does not destroy the fabrics of societies, slitting them with growing inequality. It does not destroy the world by ruthless production/consumption/pollution/population. It believes in all, hopes for all, and endures all, including its opposite.

22: We mass produce petty hedonists, narcissists, and hypersexualized citizens.

"Lots of people lie about their accomplishments and end up like shiny soap bubbles, which explode so quickly and vanish forever" (B. K.).

"As the numbers of narcissists rise dramatically, so grows the numbers of cheaters telling favorable lies about their achievements and themselves" (B. K.).

Are we becoming more mentally screwed up and self-centered?

A study published in *Psychological Science* suggested that people across the globe were becoming more individualistic over time. Individualism, as opposed to collectivism, relates to how independent and self-reliant (and self-centered) people are (July 27, 2017).

People pursuing their impulses without a sense of mutual responsibility, respect, and compassion are mass-produced nowadays. People who meet a stranger and after a very short while ask for sex are fucked up in their heads.

Due to impulse-prone individualism, availability of pornography/drugs, casual sex, and promiscuity, lots of people in our time are psychologically screwed up. They are not able to create and maintain enduring and mutual relations, as they are disturbed/infantilized in this very essential area.

What is self-focused behavior? It is the focus of conscious attention on oneself and one's thoughts, needs, desires, and emotions. An excess of self-focus has been associated with the development of or having heightened vulnerability to several mental health disorders such as alcohol abuse, depression, and anxiety disorders.

I have met lots of self-centered hedonists in my life who focus was on enjoying life, often in a self-destructive manner. They travelled often to sex paradises, where they had intercourse with young prostitutes (sometimes with children), ate and drank much, and lived cheaply. This was the peak of their lives. This self-indulgent lifestyle was their highest ambition in their lives.

A friend of mine described an old lonely man who had lived like this, without any responsibility for others than himself and his pleasures. He lived like this all his grownup life, ending up being useless, isolated, lonely, and immersed in his memories.

> E. was once a sailor in the southern seas. There he stood in 1958, with his strong bare body, his open/curious and adventurous mind, and hairy chest. With his desire to master life, to describe life, to drink life, to understand life. A lust as strong as thirst felt by wanderers without water in the desert, but with no direction beyond his own self focus.
>
> E. traveled the world. He was given crocodile skins and a hundred elephants, Spanish souvenirs, African masks and Arabic coffee. Youth, hope, joy, sex, dance, songs, laughter, camaraderie, beloved Jerusalem, but without commitments beside himself. Everything around is carefully pinned to the walls of his flat, standing on the shelves, kept in the frames, archived in the folders. Like dried summer flowers in an old insignificant book.

Time is rough and never young. Time has thrown E. into 2021.

> In the cold February light, in foreign, chilly Copenhagen, E. goes up and up and up... third, fourth, fifth floor. Fifth floor. Last floor. Eighty-six years. A stranger in the city and in the country, although he spent most of his life in both.
>
> He goes with daily zeal back to the endless chain dances on his TV screen, to songs that no one can remember, to faces and

names that are only in his memory, to Jerusalem in the forties,
to E.'s city that does not exist any longer. To E.'s total oblivion!

It should be clear that these self-centered people are the products of our modern societies, which foolishly put too much stress on whatever one considers as self-realization rather than on our responsibilities for the common good. All the people who were seduced to follow this track were burdened by anxiety with committing themselves to something greater than their petty lives and became slaves to their whims and lusts—a rather poor substitute for a more life-affirming and fulfilling life.

They did not learn the lesson that if you aim at enjoying life as a purpose in itself without dealing with *Tikkun olam/adam* (the improvement of man and our world), you miss the opportunity to pursue our true destiny.

Regarding the modern narcissistic epidemic, an extreme form of hedonism.

As our self-indulgent modern society keeps producing narcissists, all deluding themselves regarding their great qualities, they have also become skillful and boastful liars, selling fairy tales about their performances and persons in a free-market style of overblown image/success.

I have always suspected that narcissism is basically being nourished by a combination of inferior self-value, delusions of grandeur, and infantile personality. From this perspective, narcissism is not real self-love, but false, distorted self-love.

Is narcissism in the world increasing?

Narcissism is increasing in modern western societies, and this has been referred to as a "narcissism epidemic." The endorsement rate for the statement "I am an important person" has increased from 12 percent in 1963 to 77–80 percent in 1992 in adolescents (January 24, 2018).

Jean Twenge, professor of psychology at San Diego University, has argued strongly that narcissism is on the rise. She has even written a book about it, *The Narcissism Epidemic*. Her research seems to show that American college students have become more narcissistic. Is there an epidemic of narcissism today? San Diego State psychologist Jean Twenge tells us that yes, indeed, there's ample evidence to show that we're living in a culture of escalating narcissism (July 8, 2016).

Is there an increase in adolescent mood disorders, self-harm, and suicide since 2010 in the USA, UK, and other English-speaking countries?

Is there an increase in adolescent mood disorders, self-harm, and suicide since 2010 in the USA and UK? Several studies have documented increases in adolescent loneliness and depression in the US, UK, and Canada after 2012, but it is unknown whether these trends appear worldwide or whether they are linked to factors such as economic conditions, technology use, or changes in family size (J. Haidt and J. Twenge, 2021).

Hyper sexuality/promiscuity

The west may invest in its own decline by mass-producing androgynous humans, people having the characteristics or nature of both male and female androgynous heroines. Androgynous people, polygamy, open marriages with sexual relations do make people focus on their sexual organs and gender, but do not help people to transcend their lives beyond being transient dust. When sexual/ gender preferences become the central theme/meaning in people lives, it indicates strongly a degeneration of priorities in the society, regardless of the claims for the defense of human rights. And this is a growing concern regarding the cohesive/ meaning bearing future of the west!

Is being hypersexualized and promiscuous instilled in our brains? If this is the case and people can't resist it, it become obvious that it is a kind of compulsion/ obsession/flight from personal problems/turmoil and free will, draining the afflicted of energy and motivation to accomplish much more desires.

Sexuality is also part of the narcissistic culture in which we live. Identity through sexuality and "lifestyle" is what many try to attain. Yet it is but a transient status symbol for the many, as it is self-focused, a commercial-driven trend promoted by both commercial interests and mass media/ social media. Promiscuity thus becomes not just a lifestyle—often devoid of intimacy, devotion, responsibility, support, and mutuality—but what gives life meaning life for many common people without higher motives in their lives.

Research tell us more regarding this affliction among people. Women, far from being naturally monogamous, are, like men, naturally promiscuous. Biologists believe that women are genetically programmed to have sex with several different men in order to increase the chances of healthy children with the greatest likelihood of survival (September 2, 2000).

What causes a woman to be promiscuous? Alcohol or drug abuse problems;

mental health conditions such as a mood disorder (such as depression or anxiety) or a sex/gambling addiction; family conflicts or family members with problems such as addiction; a history of physical or sexual abuse (February 7, 2020).

A number of studies have found a connection between self-esteem and teen sexual activity. For example, one early study found that girls who reported being sexually active had lower scores on measures of self-esteem.

Intelligence and sublimation (family, ideals, higher purpose) teach us to be less sexually impulsive. The hectic atmosphere of a big city and loneliness-inducing relations also contribute to the growth of compulsory-prone promiscuity.

An experiment with rats under conditions that could resemble a mega city also showed the widespread of promiscuity. That many people's sexual behavior under big city conditions can remind us of an experiment on rats tells us how easily people can become manipulated by following their impulses.

23. Our polluted lifestyle results in growing degenerative, autoimmune ailments; mental suffering; and humans being dumbed down.

Neurological disorders are now the leading source of disability globally, and the fastest growing neurological disorder in the world is Parkinson's disease. From 1990 to 2015, the number of people with Parkinson's disease doubled to over six million. In the US, the number of people with Parkinson's has increased 35 percent in the last ten years (2010–2020), and the estimates are that in the next twenty-five years it will double again.

Most cases of Parkinson's disease are considered idiopathic: they lack a clear cause. Yet researchers increasingly believe that one factor is environmental exposure to trichloroethylene (TCE), a chemical compound used in industrial degreasing, dry cleaning, and household products such as some shoe polishes and carpet cleaners. TCE is a carcinogen linked to renal cell carcinoma; cancers of the cervix, liver, biliary passages, lymphatic system, and male breast tissue; and fetal cardiac defects, among other effects. Its known relationship to Parkinson's may often be overlooked due to the fact that exposure to TCE can predate the disease's onset by decades. While some people exposed may sicken quickly, others may unknowingly work or live on contaminated sites for most of their lives before developing symptoms of Parkinson's.

It is almost certain that many other growing neurological and mental disabilities like autism, Down's syndrome, mental retardations, cancer and other autoimmune ailments, and the global fall of human intelligence are connected with our chemically polluted environment.

Is there a clear relationship between gestational exposure to air pollution and cognitive development in children? A study finds that exposure to fine particulate matter in the first years of life is associated with poorer performance in working memory and executive attention. A growing body of research suggests that exposure to air pollution in the earliest stages of life is associated with negative effects on cognitive abilities (March 23, 2019). Air pollution could cause a significant reduction in intelligence, a major study says. Research finds that long-term exposure to air pollution impeded people's performance in both verbal and math tests (August 28, 2018).

The immense consumption of Junk food causing immense obesity is associated with the dumbing down of the masses. Rising temperatures in our world affects not only cognitive sharpness but also the level of aggressive behavior.

How does extreme heat from climate change distort human behavior? As temperatures rise, violence and aggression also go up, while focus and productivity decline. Intense heat can affect human behavior, from making it hard to focus to prompting aggression. Scorching temperatures have been causing workers to make mistakes or even faint near the dangerous machinery. Physiologically, people's bodies aren't built to handle heat beyond wet bulb temperatures—a combined measure of heat and humidity—of around 35 °Celsius, or about 95 °Fahrenheit (SN, August 5, 2020). Mounting evidence shows that when heat taxes people's bodies, their performance on various tasks, as well as overall coping mechanisms, also suffer. Researchers have linked extreme heat to increased aggression, lower cognitive ability, and lost productivity.

It is obvious that both the physical and mental pollution in our time has a huge impact on our physical/mental health condition. Alarmingly growing numbers of people in the world are suffering from mental instability and disturbances. According to data from the IHME's Global Burden of Disease, about 13 percent of the global population—some 971 million people—suffer from some kind of mental disorder. Dementia is the fastest-growing mental illness.

Antibiotics in early life could lead to brain disorders

Exposure to antibiotics in utero or after birth could lead to brain disorders in later childhood, says a Rutgers researcher(12). Antibiotic exposure early in life could alter human brain development in areas responsible for cognitive and emotional functions. The laboratory study, published in the journal *iScience*, suggests that penicillin changes the microbiome—the trillions of beneficial microorganisms that live in and on our bodies—as well as gene expression, which allows cells to respond to its changing environment in key areas of the developing brain. The findings suggest reducing widespread antibiotic use or using alternatives when possible to prevent neurodevelopment problems.

Penicillin and related medicines (like ampicillin and amoxicillin) are the most widely used antibiotics for children worldwide. In the United States, the average child receives nearly three courses of antibiotics before the age of two. Similar or greater exposure rates occur in many other countries.

Growing mental problems due to mentally polluting lifestyle?

The National Survey of Drug Use and Health (NSDUH) shows an increase in major depressive episodes, as is noted in the article in *Journal of Abnormal Psychology* and elsewhere. In addition, a study about children and adolescents who are eligible for SSI by virtue of mental disabilities found an increase in most disorders from 2004 through 2013. And CDC surveillance during 1994–2011 has shown the prevalence of mental health conditions to be increasing. Perhaps most significant is the documented rise in adolescent and young adult (13–25) suicide rates from 8.9 to 11.6 per 100,000, a 30 percent increase since the turn of the twenty-first century.

Our body and mind polluting lifestyle has resulted in obesity epidemics in the world.

Overweight and reduction of intelligence are intercorrelated.

Excessive body fat around the middle linked to smaller brain size, study finds (January 9, 2019, American Academy of Neurology). Carrying extra body fat, especially around the middle, may be linked to brain shrinkage, according to a research. For the study, researchers determined obesity by measuring body mass index (BMI) and waist-to-hip ratio in study participants, and found those with higher ratios of both measures had the lowest brain volume.

Illiteracy

Whatever incredibly dumb things humanity got up to in the twentieth century (and there were, as you know, some doozies), we all had at least one thing to crow about: as measured by IQ tests, humans were at least steadily getting smarter. The steady uptick in average IQ scores is known as the Flynn effect, and it lasted for decades. Basically, wherever scientists looked they found a rise of intelligence of about three IQ points per decade.

But recent research has worrying news: this trend appears to be reversing.

Humanity is now officially getting dumber.

It probably shouldn't worry us if some pocket of the population saw a decline in IQ as things like education and diet affect IQ and these factors can vary from one group or time to another. But according to this new study, it doesn't appear to be some small segment of the population whose IQ is going down. It appears to be the entire nation of Norway.

When scientists from Norway's Ragnar Frisch Centre for Economic Research analyzed some 730,000 IQ tests given to Norwegian men before their compulsory military service from 1970 to 2009, they found that average IQ scores were actually sinking. And not just by some miniscule amount. Each generation of Norwegian men appear to be getting around seven IQ points dumber.

And as PsyBlog points out, this isn't even the first study to find that the Flynn effect has reversed, though it may be the most convincing to date.

That's pretty horrifying news for fans of progress, but it also begs one incredibly important question: Why? What's causing IQ scores to start heading in the wrong direction?

You might first wonder if it's genetic. Maybe some change in the makeup of a particular group being studied has caused the decline (crudely, you could call this the "dumb people have more babies" hypothesis). But that seems to be ruled out by the new research, which shows that even within single families IQ has declined. In other words, we have started building a more folly-inducing environment.

So we know that the culprits are both nurture and nature (consumerism, self-indulgence, migration, pollution), but scientists are still baffled as to what exact aspect of modern life is driving the decline. Some have proposed that our tech obsession might be to blame, but as the decline started in the 1970s, well before everyone spent their days staring at screens, that can't be the whole story. Other proposed explanations are unhealthy modern diets, increasingly trashy media, or a decline in the quality of schooling or the prevalence of reading and a polluted environment.

The bottom line, however, is that the cause of the decline remains either a mystery or multifactorial, as I pointed out. Whatever it turns out to be, however, we should all probably start worrying about what our sedentary, screen addicted, junk food munching, mind and body polluting lifestyles might be doing to our brains.

BENJAMIN KATZ

How does pollution affect intelligence?

A research conducted by scientists from the US, Mexico, and Canada has determined that the typical air pollution in large cities may impair short-term memory, intelligence quotient (IQ), and cerebral metabolic rates. Children particularly are at increased risk for these adverse effects on the brain.

Is air pollution a global issue?

One of the most significant effects of air pollution is on climate change, particularly global warming. As a result of the growing worldwide consumption of fossil fuels, carbon dioxide levels in the atmosphere have increased steadily since 1900, and the rate of increase is accelerating.

Will pollution create a downward spiral of folly?

If our society is being dumb for not taking more urgent action about human pollution and human pollution can make us dumber, will we get to a point to where we can't even understand how stupid we are becoming? A new research paper posted on EarthArXiv suggests that rising carbon dioxide (CO_2) levels in the air will damage our cognitive abilities. That means thinking abilities. CO_2 levels in our atmosphere have hit a record high in the last years. CO_2 levels have increased much more rapidly over the past century, consistent with the thought that human activity such as fossil fuel use has been the main contributors to this rise.

Researchers found studies that showed the relationship between CO_2 levels and thinking ability, and created a predictive model of what might happen over time, given current trends in fossil fuel pollution. They drew from studies that revealed what happened in settings where CO_2 levels increased. In such school environments, student attention, vigilance, and memory decreased. In such work settings, decision-making, strategic, and crisis response abilities declined.

While these observations were associations that don't necessarily prove cause and effect, there are scientific explanations why higher CO_2 levels may affect thinking. The alveoli in your lungs exchange oxygen that's in the air you inhale with the carbon dioxide that's in the blood. If the air that you breathe has more CO_2 and less oxygen, then your blood won't get the same amount of oxygen. Oxygen, in turn, helps your cells function, including all those cells that are inside that hard spherical casing that sits on top of your neck. If your brain cells aren't getting the same level of oxygen, they may not function as well and more may even die. All of this could affect your cognitive function.

Studies are beginning to show exposure to various air pollutants also causes inflammation in the brain. Previous studies by others have shown that early exposure to lead disrupts brain development and increases aggressive behavior and juvenile delinquency (December 13, 2017).

Age-related macular degeneration, which results in a gradual and permanent loss of vision, may be influenced by air pollution, a new study finds. Characterized by the degeneration of the macula, a part of the retina, AMD impacts a person's ability to see with proper focus and results in vision loss over time (August 22, 2019).

Air pollution has been a focus of several studies on cognitive impairment and dementia risk. There is evidence that tiny air pollution particles can enter the brain, but at this time we can't say if they play a role in the development of dementia

Does pollution cause dementia?

Air pollution has proven to interfere in the human brain in such manner as to cause reduction of intelligence and increase in the frequency of neurological as Alzheimer and dementia.

Modern mind pollution

Mind pollution is focusing on consumption, pleasures, narcissism, and greed. In other words, dancing around the golden calf, lacking higher motives/aspirations in life besides one's current life, which dumbs us furthermore down. A human life devoid of challenging our mental and physical constraints and further evolving is not living up to our full potential. It is stagnating around mere living, pleasures, rituals, and self-deceptive convictions.

Modern mind pollution results also in so-called logical solutions to our self-created global problems, which seem sound but are in fact either deficient or useless in the long run. An example:

The bamboozle has captured modern human minds

Bamboozle means to trick or deceive someone, often by confusing them. And this is what really going on. Most people in the modern world have been bamboozled to believe and act as if they were unique individuals with ample possibilities/choices for both careers, consumption, and love relations. This is a lie. It is true that there are many more different jobs, goodies to consume, and single people in the world than ever, but this abundance of them don't make life and love easy. Many people never get the job they aspire to regardless of the slogan "If you work hard enough..." Many people buy obsessively, becoming dependent/addicted to buying and consuming. (There is a significant rise in the number of abusers of pornography, drugs, sex, and other kinds of obsessions.) And the worst of all is that many people in the modern world are disturbed mentally, being bamboozled to believe in themselves as the center of the world, without much understanding of others' needs. And they therefore fail miserably in both love relations and recreating a sustainable world. Both are relationships that need nurture, sacrifice, and endurance, but alas, many people cannot make these relations work. And as they grow older, they become mentally fossilized.

All these poor people think that they determine their lives, all the while their minds were bamboozled by other forces (commercial/political manipulators), reducing them sometimes to puppets on a string.

A very clear notion of this phenomena you could watch while walking on the streets of Copenhagen in the end of March 2021, around Easter. There were formed very long lines of people and children waiting for hours to get into shops selling clothes and other wares. A culture of compulsive shopping hedonists, of hooked consumers became visible amid the background of empty churches.

TikTok Toxicity

You've probably heard of TikTok, which is a trending social media app where people can make and share their own videos on the platform. It's entertaining, fun, and addictive. But recently I've noticed that the platform has changed. It's now filled with endless amounts of hate, toxicity, and even extreme cyberbullying. And what makes this even worse is that TikTok is filled with mostly teen creators, who are made the prime targets of bullying caused by herd mentality. It seems that TikTok is a reflection of this current generation, where people are expected to fit into a mold that is seen to be "correct," forcing a limited freedom of expression and creating barriers by gatekeeping almost everything from bandanas to anime.

24. The parrot education of children in schools.

These are some postulates from an article in *New York Times* (Jim Huylebroek, "The American Identity Crisis," July 15, 2021):

"Over the past decades, America and its allies have betrayed our values and compromised with tyrants innumerable times. But at their core, the liberal powers radiate a set of vital ideals—not just democracy and capitalism, but also feminism, multiculturalism, human rights, egalitarianism, LGBTQ rights, and the dream of racial justice. These things are all intertwined in a progressive package that puts individual dignity at the center."

But what are we to do with this patented idealistic zeal when it brings about conflicts and misery as the cases were in Iraq, Libya, and Afghanistan, where people, foreign and domestic, don't like that package and feel existentially threatened by it? Idealists are often difficult, stubborn, and better-knowing people than others, and on top of this, they can be terribly righteous.

The enemies of this package are leading a twenty-first-century *Kulturkampf* against women's rights, gay rights, minority rights, individual dignity—the whole progressive package, the writer claims brazenly. So he concluded, as a dummy product of American school, that this is a cultural war, not traditional power rivalry. Each civilization is thus trying to attract believers to its own vision. Therefore it matters tremendously how we show up in the world, albeit him.

This explains why we dumb down children in school, having patent on both progress and the right ideals, demonizing the other parts on this ground and keeping the circle of human animosity and lack of cooperation on essential areas, repeating itself by automatic brainwashed people who had learned this crap in their schools as parrots.

The story "The Fire" (Sufi tale) tells us partly what is wrong with our education and teaching of children and grownups.

In this story, "The Making of the Fire," the fire is a metaphor for the uniting, life-granting force in the realm of different religious faiths/ideologies in the human world. It is told about different tribes of people learning the secret of making fire, but each one of them attribute their own specific ritual or ceremony to the process. One day, a wise man and his little band of disciples were traveling through the lands of these tribes. The disciples were amazed at the variety of rituals they had encountered. One disciple commented, "But all these procedures are in fact related to making of fire, nothing else. We should reform these people!" Their teacher replied, "Very well. We shall restart our journey. By the end of it, those who survive will know the real problems and how to approach them."

So they went from tribe to tribe trying to convince them that all their different and varied rituals were essentially on making a fire. The priests and the establishments in these tribes protested violently, killing some disciples they considered dangerous to their faith and social coherence. When they ended their journey, their teacher told the survivors, "You have to learn how to teach, for man does not want to be taught. First, you have to teach people how to learn. And before that you have to teach them that there is still something to be learned. They imagine that they are ready to learn. But they want to learn what they imagine is to be learned, not what they have first to learn. When you have learned all this, then you can devise the way to teach" (Idris Shah, *Tales of the Dervishes*).

This knowledge of tending the fire (metaphorically) is available to many of us, but it is being widely used as a secondary aspect of faith or ideology. The capacity to teach it contextually to children/grownups is often lacking in both teachers and society.

On education and mind pollution

We've bought into the idea that education is about training and "success" rather than learning to think critically and challenge outdated convictions and constructs. We should not forget that the true purpose of education is to make open, searching, and challenging minds, not careers. A culture that does not grasp this, which fails to understand that the measure of a civilization is its compassion and free-thinking minds, not its speed or ability to consume, condemns itself to decline.

Our western education produces the elite of tomorrow and imprint in them the following characteristics: being intelligent, studious, humanistic, morally principled, righteous, rigid, often judgmental and partly self-controlled and controlling, and very hypocritical as to be able to think long-term life-affirmingly, contextually, creatively, and out of the mainstream box.

It promotes a kind of cognitive/mental schizophrenia. This hidden schizophrenia can be seen clearly in the world of reality, with its often many contradictions between what we preach and how we really behave in the world.

25. Greed as the cancer of modern life.

"Money obsession—greed, in other words—is just a projection of who we are. Many people wish to be appreciative of life/the world, but only few dare challenge human greed, which undermines this appreciation" (B. K.).

Bitcoin is a virtual coin, and selling/buying it is a big business. In order to make a transaction, lots of super powerful computers around the world are involved in a kind of competition to formalize and finalize the deal. This is called mining, and the winner takes it all. The other super computers just wasted a lot of energy in the form of electricity.

Looking at the ecological burden this huge energy consumption entails, it is obvious that lots of people making their living on it are greedy, insane. In one day of mining, this network of computers can use energy equivalent to 240 million freezers being on all day long. Bitcoin alone uses as much electricity as Norway or Argentina consumes a whole year. It is a criminal act of abusing resources in a time of great ecological perils, yet for the time being greed/profit seems to have the upper hand in the minds of many people.

Greed is a human phenomenon. Religious mores fight greed, as was the case with Moses and the golden calf. But religion and religious people are infected by it too, since it is deeply entrenched in the human psyche. Our immense focus on our self-interests (greed) had been illuminated by a Russian observer, Maxim Trudolyubov, who stated in an interview with *Weekendavisen* (August 25, 2017) that the current geopolitical struggle is a very cynical, calculating game without any place for values or consideration of what is good to the world.

There is not a single strong player in the power game that thinks of what is good for the human world. Russia is concerned with Russia, and so is the USA, China, and other powerful nations. This observation is also true for modern humans as individuals. High-minded talk of public service in politics often conceals greed and self-aggrandizement as the norm. Other animals also promote their self-interest as a survival tool, but humans promote it often as reckless greed, which makes them harm their own long-term self-interest and other species on earth.

An old story touches more forcefully our propensity for greed. It is the story of King Midas. He had a lot of gold, and the more he had the more he wanted. He store all the gold in his vaults and used to spend time every day counting it. One day while he was counting, a stranger came from nowhere and said he would grant him a wish. The king was delighted and said, "I would like everything I touch to turn to gold." The stranger asked the king, "Are you sure?" The king replied, "Yes." So the stranger said, "Starting tomorrow morning, with the sun rays, you will get the golden touch." The king thought he must be dreaming, this couldn't be true. But the next day when he woke up, he touched the bed, his clothes, and everything turned to gold. He looked out of the window and saw his daughter playing in the garden. He decided to give her a surprise and thought she would be happy. But before he went to the garden he decided to read a book. The moment he touched it, it turned into gold and he couldn't read it. Then he sat to have breakfast, and the moment he touched the fruit and the glass of water they turned into gold. He was getting hungry and he said to himself, "I can't eat and drink gold." Just about that time his daughter came running, and he hugged her and she turned into a gold statue. There were no more smiles left. The king bowed his head and started crying. The stranger who gave the wish came again and asked the king if he was happy with his golden touch. The king said he was the most miserable man. The stranger asked, "What would you rather have, your food and loving daughter or lumps of gold and her golden statue?" The king cried and asked for forgiveness. He said, "I will give up all my gold. Please give me my daughter back, because without her I have lost everything worth having." The stranger said to the king, "You have become wiser than before," and he reversed the spell. He got his daughter back in his arms and the king learned a lesson that he never forgot for the rest of his life.

What is the moral of the story?

1. Distorted values lead to tragedy.

2. Sometimes, getting what you want may be a bigger tragedy than not getting what you want.

3. Unlike the game of soccer where players can be substituted, the game of life allows no substitutions or replays. We may not get a second chance to reverse our tragedies as the king did.

Why does a person become greedy?

In many ways, greed is foremost a matter of the heart, of our inner lives. Greed is not merely caring about money and possessions, but caring too much about them. The greedy person is too attached to his things and his money, or he desires more money and more things in an excessive way (December 11, 2010). In the same manner, a society and a civilization become greedy.

Greed eats up a person so that he wastes away due to the heat of the bad traits it makes one develop, such as selfishness, anger, jealousy, and unhealthy competition. It sucks up every strand of happiness, and results in early death. The same refers to societies and our civilization, where the promotion of greed leads to self-destruction.

26. Talking to the wall: baseless hopes and compulsory rituals.

About hopeless hope as hype: You should learn history a bit better, as well as the history of our Earth to understand the role of hope in our minds. Hope is often hype, but it gives no guarantee against human tragedies/sufferings, or even the disappearance of species—like us, for example—from the face of Earth.

Why do people exhibit bad judgement rather often in the face of adversity, combined with over-optimism (hopeless hopes)? Is it a kind of mental-survival boosting mechanism, devoid of rationality or/and mental deficiency, to understand a complex dangerous situation? Why does almost everyone believe in an afterlife, even atheists, with ample of proof against it?

Most people curiously hold ideas about life after death, suggesting there is more to it than religion, fear of the void, or an inability to imagine not existing at all. Can it be that there is a mental deficiency built up in many of us that blocks comprehension of dangerous/troubling facts? This mental state shows itself also in our future prospects as a civilization, as many people cannot tolerate the idea that we destroy our own life conditions and thereby ourselves to a point of no return. So they will keep being hopelessly hopeful regarding our capacity to fix any wrongdoing regardless of the odds. This resembles the magical thinking of a child!

Such a similar thinking is expressed in the following joke. An eighty-year-old man prays three times a day in front of the Wailing Wall in Jerusalem. Asked by someone who noticed his peculiar habit why he prays three times every single day at the Wailing Wall instead of only once each day, he explains, "I do it for three good reasons. In the morning I pray for peace to reside on this battered world. In the afternoon I pray that a cure against cancer will be found. And in the evening I pray that understanding and compassion will prevail among sons of God."

"How long have you been praying like this?" he is asked.

"For more than twenty years," he replies meekly.

"And how does it feel doing such a thing for such a long time?"

"It feels like talking to a wall!"

Magical semi-religious and religious thinking is imbued in most of us, and may give transcendental meaning and significance to their believers, but also escapism, passivity, and self-delusion as well. Compared with people who believe utterly in "you live only once" and try to relate to our lives factually—and may therefore be seduced to be hedonistic, greedy, or self- centered—religious believers may clearly have the advantage of being driven by higher motives such as altruistic behavior, compassion, and thus greater peace of mind.

Yet this semi-religious, religious, and magic thinking is, as I pointed out before, also used by lots of people in order to avoid personal responsibility for their own lives, and manipulate others into passivity and dumbness.

27. Fatalism.

The crooked branch

A little boy climbed up an old and crooked fruit tree. He had started eating the sun-ripened, juicy fruit when he suddenly noticed a large crooked branch that looked like it was about to crack. Carefully he crawled to the branch to correct it, but along the way he lost his balance and fell to the ground. His father came running out of the house to the crying boy. When it dawned on him that the boy had not been seriously injured, he asked, "How did you fall from the tree? You are otherwise good at climbing, my boy!"

"I was trying to straighten the crooked branch," said the boy, sobbing a bit.

"Do you have soil on your head?" his father snarled. "Are you trying to straighten a crooked branch when the whole world is crooked and out of order?" (Arabic narrative).

Is the world crooked and skewed? Or is it rather our vision that sometimes makes us see its diversity/contradictions as skewed?

A good, sick, friend wrote to me: "The tree can live beautifully with its crooked branch just like people with their ailments and diseases. I'm now looking out of my window here in September. In the spring, my husband laid a little sunflower seed. Fourteen days ago there was a several meters high, erect and green cane with leaves, and at the top shone a sunflower flower that was beautiful as the sun. Now it has bowed its head, but is still beautiful. It does not survive the winter, but new seeds come. It has stood in the same place all its life and has seen nothing but life from our south wall. What a life, so poor and at the same time so rich in experiences, and so much joy it has given us!"

Yes, the world is sadly skewed and brutal because it is a reflection of our conscious and longing minds, and it will surely continue to be skewed. But fortunately we can also reduce its brutality and indifference step by step by "correcting the crooked branches" that we can reach. By doing just that, we can make the world that reflects our view less skewed/brutal. We can slowly "straighten" both our skewed minds and our world by 1) pursuing an evolving and sustainable vision beyond sapiens, which 2) goal is to make us much wiser and farsighted and longer living, and much less beastlike than we are today. We are far away from thinking in this direction, thus we keep both our minds and the world-skewed and fatal prone.

28. God relations with humans: the Holy pact.

"It is easier for people to believe in God than to think contextually, factually, and critically. It is easier for people to behave normatively than to be genuine good. It is easier for people to be shortsighted good than farsighted good" (B. K.).

A senior ultra-Orthodox rabbi and spiritual leader of the United Torah Judaism party said in April 2021 that he'd prefer a government propped up by Israel's Islamist Raam party (Arabic, anti-Zionist party) than one with leftist Jewish parties, because Israeli Arab lawmakers were less likely to turn everyone secular. What is playing out in Israel is the same political fragmentation/polarization that is hobbling America: the loss of a shared national narrative to inspire and bind the country as it journeys into 2022.

Israel and America are both nations that gave birth to themselves in the name of self-proclaimed ideas and ideals. When a country tries to be a grand, aspirational human project like that, it requires sharing some very deep things, foundational principles like God, liberty and justice for all, and an animating ethos of exceptionalism. And those deep things not only have been fractured, they are actively being fractured by a polarization industry that assaults their relevance and the trust necessary for these projects to flourish. Even faith in God can't help them out of this trap, as there are so many factions and sects working God in their own way.

Their defining external threats in the second half of the twentieth century—the Cold War for America and the Arab-Israel conflict for Israel—that had a huge, binding effect on both nations, has largely disappeared, and nothing remotely as compelling has come along to cement national solidarity.

Both societies have a high density of social networks, making them increasingly hard to govern because of the way these networks have eliminated traditional gatekeepers. Getting rid of traditional gatekeepers can be good, it opens up more opportunities for others to politically engage and tell their stories. But it can also eliminate standards in ways that erode trust in the society's common ideals and values.

Both societies have had an intense, roiling experience with a highly polarizing but incredibly media savvy populist leaders ready to break all the rules and weaken the restraints of their judicial system, state bureaucracy, and traditional media unlike any leader before them. And in both countries, the majority of citizens are religious, believing in their exceptional pact with God.

Finally, huge, long-developing demographic changes have reached a tipping point in both societies. Israel's most important demographic tipping point is with its exploding ultra-Orthodox Jewish population. Israel is a nation where about half of its children—primarily ultra-Orthodox (known in Hebrew as the *Haredi* Jews), Israeli Arab, and non-Orthodox Jews living in the country's peripheries—are receiving a third-world education, and they belong to the fastest growing parts of the population.

These families (*Haredim*) have an average of seven children, and in 50 percent of their households the men do not work but instead engage in religious study, thanks to government subsidies. They do not serve in the army, and generally deprive their children of the core curriculum in math, science, computing, and reading—which is mandated by law in every developed country other than Israel—that could give them economic independence as adults and likely loosen the grip of the religious establishment on them.

Already today half of Israel's adult population is so poor that they do not pay any income tax at all. In 2017, just 20 percent of the adult population supplied 92 percent of all income tax revenue.

Similar trends of immense poverty and schisms along religious/social lines are tearing apart American society. The USA has lost its mission after WWII. At first prosperity was a good thing. It was headed in the direction of economically lifting great sections of society and more enlightened thinking toward the underprivileged. Then prosperity became the only thing, revered above all other concerns. Large sectors of society were marginalized. Regions of the country held different visions for the good life. Religion became linked with politics in a way that had been reserved for monarchies. Political divisions were magnified for the benefit of the wealthy and the polls. Social-racial progress was vilified, education neglected for the masses, jobs outsourced, pay decreased in ratio to cost of living, work demands magnified, hungry schoolchildren from one-parent homes increased, neighborhoods deteriorated, infrastructure neglected/decayed, mass transportation sidelined due to auto industry. Gun lobbies further corrupted politics and more deaths resulted, random killings increased, mental health has been poorly supported, health care in general out of reach for poorer sectors, social media replaced libraries, individual efforts became generally unrewarded unless resulting in great wealth. Systemic discrimination ruled housing, education, job markets, policing; foreign wars supported yet more acquisition of resources for the military-industrial complex; prescription drugs in lieu of a tolerable life, etc., etc., etc. And yet the dream is still possible for many to learn, work, love, find purpose, strive for improvement, and pursue happiness on a scale unknown to most of mankind, if only this society won't implode on its own thwarted dreams.

And we still don't know what is the good God's role in the following new brave world's development: automation, robotics, AI, CRISPR, research and development, cult of celebrity, disinformation. But it is an overall glimpse of many societies headed in the wrong direction, with an ecosystem that is a ticking time bomb. Are we smart enough to better our crisis-stricken societies with God or without, in a sustainable, life-affirming manner? This is the big question!

God, religion, unity, and disunity

"When the world pushes you to your knees, you're in the perfect position to pray" (Rumi).

"Then get on your legs and struggle against its pushing force" (B. K.).

"We have just enough religion to make us hate, but not enough to make us love one another" (Jonathan Swift).

Most people are mind slaves of some religion due to their childhood indoctrination, and thus, habitual mindsets. Humans are meaning-striving creatures, and the meaning they can extract out of their short lives is mostly bound to some heavenly authority with limitless power, compassion, love, and incomprehensible wisdom.

People with such mindsets cannot imagine a meaningful existence without this form of dependence/escapade. They don't know how to generate a sustaining and enduring life meaning beyond God. Being servants of God or other isms enslave their minds as to stop thinking and acting out of their "straightjacket" of constructs and convictions. It is obvious that most people need such absolute mental constructs that can function as guidelines for their aims and actions, but at what staggering price? Dietrich Bonhoeffer recorded, while awaiting execution during the Second World War, a number of his thoughts in a work we now know as *Letters and Papers from Prison*. One of these essays, titled "On Folly," records some of the problems that Bonhoeffer saw at work in Hitler's rise to power. He wrote that upon closer observation, it became apparent that every strong upsurge of power in the public sphere—be it of a political or a religious nature—infected a large part of humankind with folly. It seemed to him that under the overwhelming impact of rising power, humans were deprived of their inner independence and, more or less consciously, gave up establishing an autonomous position toward the emerging circumstances. The sobering fact is that mentally blinded people—holding to their faulted convictions and thus, repeating their failures—are also often stubborn. They come easily under a spell, blinded, misused, and abused by the powerful entity. Having thus become mindless tools, the blind persons become capable of any evil and, at the same time, incapable of seeing that it is evil, Dietrich concluded.

Faith in God does not resolve the antagonism built in our minds. Even when God leads us to a promised land, we go again and again through three recurring mental stages that encompass the essence of the contradictory human drives and aspirations. Egypt is the first stage. It is the land of slavery, where human soul and body are enslaved by wish for relative security, food, shelter, and the mere right to live. In this stage, human beings don't entertain aspirations regarding pursuing great missions/visions, although they may be badly needed since they are in pain for being enslaved.

The second stage takes place when the pain of slavery has become overwhelming due to shifting circumstances, as in Moses' tale. The pain/impending necessity motivates/forces them to desert, wandering toward a promised land where they can gain freedom. This is the stage, with all its dangers, risks, and discomforts. The desert wandering is a very demanding effort, but its promised fruits are alluring. In this travel, the slaves transform themselves slowly into real free men who can make their own free choices, believing in something greater than their petty, painful, and relatively secure slave lives.

The third stage is settling down in the Promised Land, enjoying the fruits of new freedom, and then starting slowly to corrupt this freedom with whims, greed, and folly, degenerating thus into a new form of slavery—the greedy one—by hoarding, consuming, and leading a decadent lifestyle. Is this conflict we humans exhibit unsolvable, and are we destined to go on and on to the end of our species?

Faith in God as justification to become inhuman

An acclaimed research book, *Soldiers of Evil* by Tom Segev (Jerusalem, 1987), describes SS leaders in the concentration camps in the Second World War, their backgrounds, their faiths, and their immense cruelty. All of them seemed to believe in God, and most of them came from Catholic and Protestant homes where faith was very strong. Therefore, some of them kept attending prayers in church even though the SS did not approve of it. They loved some people but hated lots of others as to consider them sub humans, destined to be destroyed. Historical experience and evidence have shown time and again that when people hate others and find some justification to view them as sub humans, they destroy them by justifying their own acts also from the viewpoint of their religion and God.

Thus, God becomes what people wish him to be—in prime cognitive-dissonance style. And since humans are far from angelic, they also use their faiths and holy convictions to justify their murderous and predatory nature.

In his book *Soldiers of Evil*, Tom Segev presents an SS document of questions and answers designed to inform the SS soldiers regarding their views of God. This document illustrates clearly why faith in God can be very divisive and counterproductive to humanity's further progress when faith comes to supplement a political power:

> Question: What is the oath an SS member makes?
> Reply: I swear for you Adolf Hitler, that I will be loyal and courageous all the time, and to my dying day I will obey your orders and the orders given to me by the officers whom you have appointed to lead me, so will God help me!
> Question: Does that mean that we believe in God?
> Reply: Yes, we believe in God. God's spirit guides us all the time.
> Question: What does SS think of people who don't believe in God?
> Reply: The SS maintains that those who don't believe in God are sick of megalomania. They are arrogant and foolish and the SS has no, whatsoever, interest in them.
> Question: Why do we believe in Germany and our Führer (Hitler)?
> Reply: As we believe in God, we believe in Germany which God had created and in Adolf Hitler which the Good God sent to us, be blessed His Grace (page 67).

29. The pursuit of happiness through infantilization.

Seeking happiness in our lives is extremely popular in many parts of the world, even though it is a futile pursuit. It is like trying to catch the wind, as happiness is not a stable state and seldom stays for long. You may be more content and fulfilled instead of this Fata Morgana pursuit if you work for the long-term good and improvement of people and mankind with the right (sane) mindset.

I have maintained for a long time that fruitful self-reflection must result in either corrections of faults and bad conduct or "I am good enough as I am," and thereafter a man should focus on bettering his world and humanity—thereby bettering himself—instead of indulging in self-focused realization. Now, studies show that spiritual self-realization aimed at being happy may result in narcissism.

A spirit of infantilism swirls around the western world and takes hold of language and behavior in various forms, and it is pushing forward this futile idea of attaining happiness and harmony. Lots of people follow this false course, being infantile in their behavior. In our western culture, we use profusely both superlatives and lies to describe people as being special and unique, all the while they are just ordinary people. This is counter-productive as it renders people passive.

Infantilism is no stranger to culture. Infantilism exists partly due to our consumer minded, self-indulgent, and ego focused individualism. Yet it is to be found at the core of the major religions, because they are built on blind faith. The believing person does not develop maturity and independence, but trusts a God who has supposedly invented things for him/her and demands doubtless faith. This childish faith element also exists regarding self-realization and happiness. Although this infantilization in pursuing happiness is different from that attributed to religiosity, it is contagious and easy to cultivate because it is an inevitable product of mass narcissism, and is spread mainly through the dominance of digital platforms and consumerism's subtle messages.

In fact, digital equipment and how it is used are designed in a way that is fundamentally adapted to children's whims and wants. Even toddlers and babies are drawn to screens, phones, moving shapes on the screen, and the dynamic colors. It does not take much more than that to capture attention, even for older ages.

Another characterization of infantilism is the tendency to attribute an authority/celebrity the power to generate great insights, magical changes, and great, pleasurable lives. Bill Gates, for example, is placed in the position of a super teacher, a kind of grandmaster who each year releases a statement for the benefit of the public, in which he takes up scores for various phenomena around the world and explains what needs improvement. Only in a childish society that tends to pursue false charms can a software man, however common, gain iconic status and be a god. But often these messages regarding attaining happiness bring forth its opposite.

In many contemporary literary works, one can find characters who seem to curl up with their bear, touching the world with a kind of artificial childlike innocence that cuddles up on itself. Their inner world is always too sensitive, too self-conscious, twisting around its weaknesses and gaze, and mostly worded with hesitation and a lack of determination.

30. The mass production of mediocre people.

"Selling us the false idea that we are unique in mass societies without performing any special feat is debilitating for the soul" (B. K.).

People in our democracies, being dumbed down by commercialism, consumerism, trivia and trifles, are steered toward mediocrity by brainwashing tyranny. It is as simple as this: Being told that they are special/unique without presenting a proper goal/challenge makes them confused, and they don't dare challenge the premises of this brainwashing system. The reason most people get stuck in mediocrity is because they refuse to think and act beyond their comfort zones. Their fear of fiasco paralyzes them to the point that they won't charge.

Being mediocre means being persons of only average quality, not very good. A person that is not very good at something or not very good at anything in particular. Mass societies are inclined to produce humans who don't challenge their premises, narratives, and values. When democracies began to flirt too much with commercialization of life and soul on account of their enlightenment and higher aspiring vision, the cubes were cast: the mass production of mediocre consumers, harming their own life conditions on earth.

Our institutions nurture mediocre minds and souls en masse, thereby rendering us the possibility to become both wiser and more farsighted as to realize both our evolving potential and a viable road to a sustainable world. We are not going to get more than mediocre policies, neither with Biden (2021) nor with Obama-like leaders. They talk in a civil manner but they are mediocre stamped, and their priorities for their nation and the world have not changed American society or the world into a better place to live in. As long as we don't shoulder our responsibility, by scarifying some of our lifestyle, a global sustainable vision and praxis will be daydreams.

Why will most people remain mediocre?

Many people are convinced they are incapable of achieving great things, so they aim for mediocrity. Many think, apropos the Dunnings/Kruger effect, that they are great achievers when the truth is that they are mediocre ones. Most people will never be truly exceptional for these two reasons. The pull toward mediocrity is too strong.

Why are people comfortable with mediocrity?

Living in mediocrity creates a mid-level comfort zone. Many people around the world are comfortable living in mediocrity. They pick the path taken by many and follow it till the end because it provides security. A mediocre life causes you to put aside your dreams and ambitions for solutions of comfort and security that make more sense to you.

Can modern mediocrity be interrelated with the current dramatic growth of anxiety and depression and the fall of intelligence scores?

It is, in an intricate manner. As the ideologies of commercialism and consumerism allied with self-focused individualism pushed aside the efforts to enlighten people, the outcome of this mind campaign was a growing number of narcissistic and mediocre people, as people have been dumbed down. Another effect of commercialism/consumerism combined with individualism that can be clearly observed is the dramatic rise of mental illness, lifestyle ailments (overweight and obesity), and the reduction of measured intelligence, ergo the dumbing down of the population

Is modern mediocrity thus connected with narcissism and consumerism?

Yes. When the focus of man is mainly to gratify his needs and whims, it results in both mediocrity and narcissism.

Right now, the sad geopolitical fact is that we have mediocre leaders who lack statesmanship and viable, evolving global vision/mission. They are better than oil salesmen, thugs, gangsters, sycophants, and lackeys, yet they are lower due to their incompetence, life standards, enlightenment, and mental health for everyone. They have reached their level of incompetence and cannot go any further. They control people on the street and in the universities who are content with a low-level existence and any shiny object presented to them. Their knowledge of history is shallow. They have not and will not learn from past mistakes. Our leaders are not the knights in shining armor they profess to be to help us out of our global growing predicaments. Yet the majority of voters (being mediocre) will continue to vote them in.

There is a beautiful, metaphoric Chinese story about a stonemason. When he observes a rich merchant coming into his village with his procession, he considers this man to be the mightiest man in the world, and wishes to become this powerful man instead of being a stonemason. A miracle occurs, and he becomes a rich merchant. But he soon realizes that the province administrator is mightier than him, and so he wishes to become such a mighty administrator. Soon he realizes that the emperor is stronger than him, and wishes to become the emperor. In this succession of events he wishes to become the sun, which is stronger than the emperor; the clouds, which the sun's rays cannot penetrate; and thus be stronger than the sun and the wind, which can play with the clouds as it wishes. He becomes all these things, and he is now sure that he is the mightiest force in the world. Then he, as a windy storm, blows and blows, but a cliff stands in his way and won't yield. So he concludes that the cliff is the strongest of all forces. He becomes a cliff, and in the morning, being a cliff, he hears the sounds of a mason breaking down the cliff with his hammer.

This is a metaphor for people traveling all around the world in search of their true destiny, only to find out that it was, all the time, right there where they lived. It tells about the meaningless striving for more/different power because all the while human power, in itself, is relative. They can get both power and meaning if they look closely, as it may lie in front of them, right where they had started their journey. Yet the journey—as this metaphor indicates—is necessary as a means to make them realize this. It transforms their minds with different experiences and insights but not their basic life situation. Its point is humbleness/humility, and the meaning derived by small acts that they attain. But the price they pay for their humbleness is high; they stop challenging their own old prison by accepting that their destiny lies within its confines. This is a mediocre life attitude!

And these confines become a mental straightjacket when they also become afraid of social isolation, as the following Sufi story tells.

> A man interpreted a message from the teacher of Moses to mean that on a certain date the water in the world would disappear. It would then be renewed with different water, which would drive all men who drank of it crazy. This man collected the old water and stored it. When the day came, the event took place as anticipated. The old water dried out, and after some time, new water started flowing. The man walked among the people he had known and realized that they thought and talked in an entirely different way than before. When he tried to talk to them, he realized that they considered him to be crazy. For a while, he drank his own stored water, but since he could not bear his loneliness, behaving and thinking in a different way from everyone else, he started drinking their water. And he forgot all about his own store of special water, and his fellowmen looked at him as a madman who had been restored to sanity (*Tales of the Dervishes*).

The human race is basically a meme-prone herd. Here we are, potentially unique, and yet we easily conform and become uniform—to put it simply -as a herd. Once we concede to the herd mentality, we can be controlled and directed by a tiny few. And we are being controlled. We are terrified of being ignored, stigmatized, and excluded, which makes it easy to form us into such a conforming herd. Yet our society's propogandists tell us that we are unique.

This fear of being excluded is wired into our brains for a good reason: our survival, physical as well as mental, was dependent on us being together with others and getting their protection, help, and support by being and behaving like them for most of our history. Groupthink is the road drum of mediocrity, and describes a psychological phenomenon where peer pressure prevents critical thought. Eight symptoms characterize groupthink: (1) the illusion of invincibility; (2) stereotypes ("us and them"); (3) rationalization ("we can always explain our failings and shortcomings"); (4) the belief in moral superiority; (5) censoring thought; (6) the illusion of unanimity; (7) pressure on deviants; and (8) the fear of exclusion keeping disruptive ideas out and cementing the cohesion of the group. Once people identify themselves with a reference group (it often happens in childhood), they acquire its shadowy side: semi-omnipotent or vainglorious attitudes combined with prejudice toward inferior outsiders, and mediocrity/conformity. Cozy, conforming, complacent, being consumption minded, and avoiding conflicts with the group = mediocrity, which constitutes a threat to our common future.

Sapiens thus act as mediocre as Maj fly:

To sum up the state of the common man poetically:
Every day, each and every day
I'll escape my destiny; a Mayfly
I plan great deeds and to myself say:
Tomorrow these deeds, realize I may.
But then I dream of a sunny bay
where I all day long seduce and play,
where temptations my heart do sway
and so goes my willpower astray.
Although I plead: Oh with me stay!
Although for strength to God I pray
I wake up to a new day as I am: a Mayfly

31. Replication of Rome's patent for peace inside its walls—bread, circuses, self-indulgence—while waging endless wars.

This phrase originates from Rome in *Satire X* of the Roman satirical poet Juvenal. Roman politicians passed laws in 140 CE to keep the votes of poorer citizens by introducing a grain dole. Giving out cheap food and entertainment, "bread and circuses," became the most effective way to rise to power.

What was the bread and circuses policy? How did Roman politicians benefit from it? It was basically a policy that provided free wheat to citizens and also entertained the citizens with free circuses, like the gladiator events, self-indulgence, and sexual promiscuity. They benefited because it kept the people satiated, and thus kept the politicians in the people's good graces.

Rome's policy has been copied and adapted by many modern nations, including the concept of keeping international conflicts simmering. Getting their hands on the resources of others while keeping order at home by supplying the citizens with food, entertainment, and self-indulgent services. The citizens are kept pacified and passive, even when they voice their dissatisfaction of their incompetent politicians through their right to express the so-called teethless free speech.

32. Our denial trap: Denying disturbing problems.

There are three denials that many people are experts in and which will determine our future: denial of reality, denial of our limitations, and denial of the huge sacrifices needed in order to alter our self-destructive global course.

Human nature is not made to deal with obstacles that do not directly and immediately affect our life, like climate change. Even when people are confronted by facts on the ground that their lifestyle destroys and kills them/other animals, or that climate change is directly making people's lives less safe, and in some cases deadly, they are not inclined to act as long as they are not directly harmed.

The story of two Jewish disciples' view of reality indirectly shows one way of denying reality. The place is Poland, a big market day on the outskirts of some city. A myriad of people are gathered in the marketplace. Farmers with their wives, and Jews looking for good, cheap goods while admiring the munching horses and cows. Here, many years ago, two Jewish students from two small villages met. They greeted each other, but were quickly entangled in a heated discussion about which of their respective rabbis had performed the most and greatest miracles. After a long and impressive word duel, one disciple said, "Listen to this example! We were standing inside the synagogue in deep prayer, and suddenly our rabbi shouted, 'God help us, my sheep! I see a house in Warsaw full of Jews burning down right now!'" "And," the disciple added, smiling slyly, "do you hopefully know that Warsaw is 400 kilometers away from our village?" "Well," said the other disciple, "and what happened to the house? Did it burn down then?" "No," replied the first, annoyed, "but that's not the point either. The important thing is that my rabbi could see so far."

Sweet tale? Many Jewish disciples and their rabbis died in the gas chambers of Nazi Germany, mainly because they denied reality regarding the Nazis' atrocities, and believing that God would not accept these atrocities and therefore would come to their rescue.

The same denial of ill effects of capitalism/consumerism/liberalism can be detected in the blind faith in them nurtured by countless people. What may trigger coordinated action against such denial can sometimes be great sufferings, as the case was with the Holocaust, making the Jews determined to have their own state. The same may happen with humanity. It may one day sacrifice western lifestyle because it is forced to do so after great global calamities and sufferings.

We are capable of taking action to avert menaces, but often only when we are on the edge of disaster with no other choice. Otherwise, people will fight a needed change if it affects their privileges and comfort zone. It is very easy to set people against you when you come up with some new ideas that endanger their comfort zone.

The COVID-19 epidemic killed more than five million people and harmed many more, reduced the global economy by 5 percent, 140 million jobs were lost, the number of extremely poor people in the world rose by 95 million, and 80 million more people are undernourished. Here we rose to the challenge, but very clumsily as this was only a warning of more deadly epidemics to come.

As our laboratories churn out modified potentially lethal viruses, we encroach on nature, where there are many deadly viruses awaiting us. And we live very densely in big cities, travel a great deal, and ceaselessly transform goodies, services, and new lethal viruses, fungus, and bacteria, while our number grew from 2 billion in 1950 to 8 billion in 2021. This cocktail of circumstances invite viruses to the great kill. Do we change our lifestyle knowing how vulnerable we have become as a civilization? Not that much as to prevent another dangerous virus from spreading havoc in societies.

All these self-created problems endangering our very existence are hard to deal with effectively as long we are so good at denying their impact on our health and life.

33. Mumbo jumbo, junk talk in mass media aimed at pacifying the masses.

"Mass media keeping people's minds focused on fleeting, shifting stimuli/ trivia is a politically directed mendacity. Telling people that they are unique without achieving anything worthwhile is a political directed audacity.

Splitting hairs or 'one cannot see the forest 'cause of the trees' has been elevated by the mass media into an opium for the masses" (B. K.).

George Carlin quotes on politics: Keep in mind, the news media are not independent; they are a sort of bulletin board and public relations firm for the ruling class—the people who run things.... If the parent corporation doesn't want you to know something, it won't be on the news. Period.

Lots of people can't distinguish between essential and secondary and nonessential things. They can't even distinguish between what their desires, drives, and compulsions make them focus on, on the one hand, and what is existentially essential for them to deal with in their lives on the other hand. For this reason the matter of gender fluidity has become a central issue for many, not realizing that it can be a matter dealt with privately and by law, and that there are much more urgent and higher purposes and priorities in life.

If the essence of human life is to learn to live in peace with himself and then do good in his world and evolve further, what may distract us from this ultimate goal is compulsions like gender fluidity. Gender fluidity, LGBTQ, being important political issues in a world undergoing immense ecological crisis and lack of sustainability is clearly a sign of decadence promoted by the mass media.

In this respect, our view of economic growth by becoming more effective is also a kind of mumbo-jumbo talk promoted by our economists. They claimed for a long time that increase in efficiency, such as burning coal more efficiently, was a good thing. But as the efficiency of burning coal grew, so did the demand for more and more coal, and far beyond the initial quantity used before using more efficient measures. This paradox is visible in the history of technological improvements of all kinds. Better cars, miles per gallon, more miles driven. Faster computer times, more time spent on computers. Cheaper energy to fly passenger flights, more flights and more passengers. Thus, the idea that efficiency is always good is redundant. Efficiency has to be put against the health of our planet, biosphere, and human/animals all bound together. Efficiency resulting in growing greed, consumption, and pollution is thus bad and should be shunned. This means also that the way many economists think—viewing growing efficiency and economic growth as preconditions for granting good life for humans and humanity—is perverted, as it supports scrupulous policies that bring humanity down. In other words, less efficient, more robust, and enduring goods, and maybe reduced levels of production, can serve us better than the current policy argued by economists.

And in this connection comes up the political shortsighted culture with its mumbo-jumbo junk talk on account of long-term, sustainable, and life-affirming strategies. Everybody know politics is a dirty business. Yet our greatest national heroes have always been politicians. Maybe there's a reason for that. Maybe it takes a certain kind of person to get down in the mud and come out with the bricks of statecraft.

Some people claim that politics is not really a dirty business but it is incredibly competitive. Politicians are constantly preparing for the next election or power shift, and in order to do this they have to keep special interests groups content/happy. If everybody knew what they wanted and how much time, money, and energy should be spent on every project, then politics would be clean and really easy. As it stands, no politician can make everyone happy. And each time they make someone unhappy, this person/group will vote for someone else or protest violently. You have to meet as many of the different and conflicting demands of the general public as you can, while your every decision is broadcast on Twitter to be twisted and used by your opponents to make them seem better suited to your job. Add in the need to keep long-term funding for your next campaign and your personal life being fair game for national news and you have a pressure keg of responsibility, which means most political careers end badly.

Most politicians start with the best of intentions, but with everything we expect of politicians, it is easy to see why taking the easy option or making a poor decision can seem like a good idea at the time. Politicians, trying to be popular/likeable, sometimes become to the aggravations of matters beyond their own domain. Ideologies and religious aspects do play a role in dimming their wits.

So don't delude yourself. Politicians, although claiming to hold to high moral ground, are amoral, based on self- interest, muddy, dirty, tricky and shortsighted. There is no real reason to feel that there will be a focus on long-term global goals for the sake of future humanity in this opportunistic area of human shortsighted interests and power games.

34. Hyper humanism with enshrined human rights but without human obligations is a disastrous thing to pursue.

What are the five principles of humanism?

We believe in the common moral decencies: altruism, integrity, honesty, truthfulness, responsibility. Humanist ethics is amenable to critical, rational guidance, or is it?

Do humanists believe in right and wrong?

Humanists believe that human experience and rational thinking provide the only source of both knowledge and a moral code to live by. Humanism is a democratic and ethical life stance that affirms that human beings have the right and responsibility to give meaning and shape to their own lives. Humanists have a duty of care to all of humanity, including future generations (mission). Humanists believe that morality is an intrinsic part of human nature, based on understanding and a concern for others (connection), needing no external sanction. Humanism is rational (critical thinking).

So why does this humanism lack obligations and sanctions in the following crucial areas for our long-term survival? These are the modern mortal threats of our time: over population/consumption/production/pollution and human shortsightedness and reckless greed.

So why don't claimed humanists in our time demand from themselves and from others restraint regarding our consumption, propagation (too many people on earth, stripping its resources), pollution, and production? If they are rational thinking, how come they can't see that our current practice of humanism grants rights but doesn't demand restraint and obligations to counter the abuse of these rights?

Honestly, our practiced humanism nowadays, which practically does not promote a sustainable world, is a crime against the coming generations.

As humanism is practiced nowadays—without demand for human obligation to promote our long-term survival—it is more escapism than a global solution for us.

35. Sapiens are warmongers and divisive creatures.

"All wars are a symptom of man's deficient brain as to think out of and beyond its programmed beastly repertoire. Regardless how fine, favorable, and nuanced we view our essence, our inbuilt warmongering nature will gravitate us down" (B. K.).

"In fighting and projecting force, we shortsighted creatures excel, while in living up to promises regarding climate we fail miserably. We will not be able to resolve our most crucial global predicaments as long as our nations' interests and our divisive warmongering dominate our mindsets" (B. K.).

There are many conflicts in the world, most of them don't get the media or policy attention as the wars in Iraq, Syria, Afghanistan, or Ukraine since they may not have the same geopolitical or economic importance. Yet they go on and on, lending weight to the observation that we are basically warmongers.

The toll of decades-long conflicts from Colombia to the Ogaden, from Kashmir to Western Sahara, is just as devastating for the people who live there. The number and intensity of future conflicts due to lack of water and food, climate change, and massive immigration/refugees will probably grow dramatically in this century and probably beyond as a struggle for diminishing resources will intensify.

In our time (2022), our institutions have begun to crumble under the pressure of climate change, growing populations, and dwindling resources to the point where higher concepts of law are becoming obsolete. The myriad conflicts in the Middle East, Africa, Asia, and the big powers, including the murderous criminal cartels all around the world, depict a very bleak future for humanity, full of pressure on global resources worldwide.

Can we stop being warmongers just by power of will and good intentions, when the struggle for resources and suitable places to live intensifies due to climate change? This is but a delusion!

36. The impact of being unemployed, useless, or overworked.

As the current projections for the future indicate that more and more of the work now executed by humans will be done by robots and other machines, it means that there will be billions of people pushed out of the workforce who will have to entertain themselves to give meaning and structure to their idle lives, living on some meager pension at best. You don't have to be wise to know that without some meaning conveyed by being useful in social contexts, people will become depressed, listless, and even a danger to themselves (due to frustration) and to other people's safety and the stability of future civilization.

Human beings are social animals, and need both the protection and acceptance of their fellow men. Most of us also need the feeling that we have some significance for other people by doing something that conveys respect, like work. Work is a way of conveying to people that they are useful and are paid for their efforts with both money and social recognition. Not being used at all and just living a life for one's own sake; or doing a meaningless, useless, idiotic job for the rest of one's life; or becoming a criminal in a gang is not only demoralizing but mentally degenerating as well. Total mortality among unemployed men is 2.54 times higher than among employed men. This excess mortality falls by almost 40 percent when sociodemographic background variables available at the census are simultaneously adjusted for. Nonetheless, relative excess mortality of 1.93 still remains.

Unemployment has long been associated with a significantly increased risk of death in general, particularly for low-skilled workers in the US. The risk of heart disease, the leading cause of death in the US at almost 650,000 deaths per year, has been shown to increase by 15–30 percent in men unemployed. Some previous research indicate that unemployment is correlated with worse birth outcomes, including higher rates of low birthweight and infant mortality, possibly due to stress-related endocrine system changes. Yet there are hundreds of millions of people, if not more, who end up in such a mental quagmire. On the other hand, there are hundreds of millions of workers, if not more, who work fifty-five hours a week, if not more, and are thereby are—according to solid research—at high risk of dying prematurely due to the ill effects of accumulated stress on their body/health.

In *Politiken* (13) a new study by the WHO concluded that people working over fifty-five hours a week had 35 percent higher risk of being hit by stroke and 17 percent higher risk of dying of a heart ailment than those who work thirty to forty hours a week.

A dichotomic world of billions unemployed will surely invite chaos and high criminality, even warfare, on the one hand, while the overworked in it will shorten their lives doing just this.

37. Conspicuous consumption: The imitating/competing ape.

"Take human vanity and mix it with our urge to belong to a group and at the same time be signaled out as unique, and you will know how people are being manipulated to consume excessively" (B. K.).

Conspicuous consumption is a term used to describe and explain the consumer practice of purchasing or using goods of a higher quality or in greater quantity than might be considered necessary in practical terms. Conspicuous consumption is a theory that is both economic and psychological. The economic conditions that an individual resides in can be a deciding factor as to whether a person decides to conspicuously consume goods or not. While many factors contribute to conspicuous consumption, the driving force behind such activity is the desire for the "recognition of others," as Thorstein Veblen famously stated in his book.

Man, said Thorsten Veblen (p. 31), is a creature of strong and irrational drives, credulous, untutored, and ritualistic. Veblen was a stranger to this world and an incarnated nonconformist. The world for him was uncomfortable and forbidding. He preserved his integrity at the cost of frightful solitude. And maybe therefore he could see certain things so clearly. He delved into the nature of man and his economic rites and rituals. He was seeking to penetrate the true nature of society. And in that search, through a maze of deceptions and conventions, he would have to take hints anywhere they revealed themselves: in dress, in manners, in speech, or in polite usage.

Modern man was, in Veblen's eyes, only a shade removed from our barbarian forebears, who by winning wealth by force came to establish themselves as a dignified and horrific leisure class. The discipline of savage life, he wrote, has been far the most protracted and probably the most exacting of any phase of culture in all the history of the race. Heredity dictates human nature, which must indefinitely continue to be savage.

In modern life he saw how the leisure class had refined its methods, but its aims were still the same: the predatory seizure of goods without work. Every one—workman, middle-class citizen, and capitalist—uses the conspicuous expenditure of money (indeed its conspicuous waste) to demonstrate his predatory prowess.

In his book, he followed the line of predatory nature, claiming that businessmen are virtually interested only in one thing: profit. Every economist from the days of Adam Smith made the capitalist the driving figure in the economic scheme. For better or worse, the capitalist was generally assumed to be the central generator of economic progress. But with Veblen all this was turned upside down. The businessman was still the central figure, but no longer the driving force. Now he was portrayed as the saboteur of the system.

Veblen was thinking of the kinds of overkill that were going on around him at the end of the nineteenth century. This conspicuous consumption had to combine a studious exhibition of expensiveness with a make-believe simplicity. For Veblen, what distinguished capitalist civilization in his time was that the symbolic pageant of invidious comparison replaced nearly every other kind of human relationship. He found examples for these in buying expensive dogs, in the lawns people grow that pathetically mimic aristocratic pastures, and in the lower middle-class man's decision to overwork himself so that his wife could become the bearer of symbols of "vicarious leisure." The law of conspicuous consumption makes the sensible choice impossible, let alone the feelings of being content and happy.

Whether this critique holds or not, the knowledge that we somehow are governed by deeply ingrained needs/whims, not reason, and that we can't see the difference between the painting and its price tag is not that hard to determine. Yet defenders of the system will claim that there is nothing reasonable about most of our needs, which isn't to say that needs can't be reasoned or at least argued. The defense line of the capitalist system counts on what we call life. There is nothing to do with it other than to live it. There is nothing to do with a dress other than to try it on.

But is a happy life to get whatever you wish in your childish state? Modern life isn't conducive to mental health or happiness! There is no doubt that such a lifestyle seems to pursue happiness, but its path is not self-evident. And that's another flirtatious example of why the system is so vibrant and alive. Delivering the goods on the one hand and yet seeming to dig its own grave through overconsumption and competitiveness on the other. This course is both invigorating and destructive for the human mind and environment.

The inflated human demands for more happiness have engulfed the very psychological, social, and economic cores of western civilization. Stretching from the close of World War II, it is best defined by its soaring ambitions. We had a grand vision. We didn't merely expect things to get better. We expected all social problems to be solved. In our new society, most workers would have rising incomes and stable jobs. Business cycles would disappear. Poverty, racism, and crime would recede. Compassionate governments would protect the poor, old, and unlucky. We expected almost limitless personal freedom and self-fulfillment. We not only expected these things, after a while we thought we were entitled to them as a matter of right.

In *Theory of the Leisure Class*, Thorstein Veblen argues that once an item becomes widely owned, possessing it becomes a requisite for "self-respect." People try to consume just beyond their reach so they can "outdo" those with whom they compare themselves. In such a system, frustration and overconsumption is preordained.

Does this show an innate greed characteristic for the modern mass human?

Most humans tend to focus much on the here and now and less so on the macro, evolving aspects of our further journey.

It reminds me of the following allegory. Two grains of sand were rolling over the sandbanks of the Sahara Desert when one of them turned to his partner and said, "I think we are being persecuted." Yes, in the absence of real great meaning driving billions of sand grains, lacking a binding grand mission/vision, this sense of self-importance can be generated via delusional notions. It is the human twisted mindset that demands fundamental correction. Unless we remove this mental defect from sapiens, we will keep on focusing in vain on the same micro issues, ignoring the macro problem of our species' brain/mind incapacities, and thus seeing clearly what our ultimate meaning in our journey is all about.

38. The idiocy of theorizing without possessing the capacity "to swim" (Nasiruddin).

In Yiddish there is a person called a *kibitzer*, a person who looks on and often offers unwanted advice or comments to anybody who will listen to him. He consequently lacks the capacity to live up to his own advice. It is a known phenomenon that what many people consider grants them social status is not in congruence with their abilities to resolve menacing problems and deal skillfully with one's own and others' problems and crisis. Our world is filled with people with education, expertise, and access to information. But judging from our capacity to prime our survival as persons (lots of people harm themselves), nations (ongoing conflicts due to power lust), and a civilization (destroying our life conditions on earth), we are often failing the test.

The old maxim that for most humans good intentions aren't followed by proper actions seems to be valid in a disturbing manner. In old times, it was ignorance and folly that gave nourishment to self-glorification and willingness to lend useless advice that oneself could not follow. Nowadays it seems that ignorance, folly, specialization of people in professional areas, and the brainwashing of not free at all mass media are united in causing this damage; the growing distance of what people proclaim is right to do and what they are capable of and willing to do. A striking example is that most people in my country (Denmark) wish to do much to improve their environment and fight climate change, but does not account of their life standard and style, which is immensely resource demanding and polluting.

Conclusion: If you can only see the faults, brainwashing, and suppression of free thoughts in other political orders and not in your own, you are for sure brainwashed.

39. People prostitute themselves for honor/status by losing their personal integrity.

Countless people are willing to follow dubious morality, compromising their personal integrity—many of them work in double morale prone areas—by pursuing power and influence, as the following story tells.

It is the story of a Jewish man in old Eastern Europe who comes to visit an old friend who lives in a godforsaken and poverty-stricken little town. Upon arriving at the station, he asks a passerby if he knows where his friend Moyshe lives. The passerby spits on the ground and says, "Oh, this Moyshe, this good-for-nothing idiot, lives in the next street, close to the synagogue." The guest continues his search and finds another Jew, and asks again for the whereabouts of his friend Moyshe. "Oh, this intolerable creature! This terrible being lives just across the next street." The guest cannot orient himself in the small Jewish ghetto, and asks another Jew the whereabouts of his friend. The third man looks at him and asks, "Are you friends with this overblown, pretentious, good-for-nothing being? He lives just across the corner in his little room." At last the guest finds his Moyshe, and they are very happy to see each other after many years. When they sit down, the guest asks Moyshe, "Say, my good friend, what do you do for your living?" "I have a very important job in the synagogue. I am responsible for the prayers," Moyshe says. "Do you earn much money doing this?" his guest asks. "No," says Moyshe. "I barely make a living out of it." "Do the Jews here who come to the synagogue like you very much?" "Not a bit, they hate my guts," mutters Moyshe. "So why do you do this job, my friend?" "And honor and respect do not count any longer?" Moyshe asks his guest.

Defending one's honor can result in submitting oneself to idiotic norms like risking one's life. Two Russian poets, Alexander Pushkin and Mikhail Lermontov, among many other people, died defending their honor against all odds. They were challenged by officers to a duel, where they had no living chance to survive. For them this matter of honor was instilled in their minds, so they could not see the futility of being dragged into such an unfair duel.

Something similar keeps happening between nations in our time. Not losing face, keeping up your deterrence, not giving in, keeping up vain pride among your citizens are all slogans originating from this source of defending one's honor. Lebanon, Gaza, and Iran—all on the verge of economic collapse, where citizens are plunged into merciless poverty—keep on fighting Israel due to their perceived justice and honor even when they can't win, and their people suffer immensely due to their politicians' priorities.

40. The denied, devastating envy.

There are two feelings people exert special efforts to cover over even though they are very common. Envy and the feeling that their lives had not lived up to their expectations. The day humans will have more generosity in store than envy will transform them from sapiens into higher beings.

Lust, gluttony, greed, sloth, wrath, envy, and pride are the seven deadly sins. And envy is the dumbest and most futile of them all. As an investor in these feelings, you get something out of all the deadly sins, except for envy. Being envious of someone else is pretty stupid. Wishing them badly or wishing you did as well as they did—all it does is ruin your day.

In his book *Egalitarian Envy*, Gonzalo Fernández de la Mora noted that envy is a widely denied emotion: "One may admit to pride, avarice, lust, anger, gluttony and laziness, and one may even boast of them. There is only one capital sin no one admits to: envy. This is the dark, hidden, eternally masked sin. One tries to hide it from others with multiple disguises; its symbol ought to be a mask."

Almost no one admits to envy.

Here is my friend's (Christian Ifversen) life philosophy regarding envy: It is a gift from you to yourself. A sour gift, but still a gift that you have to receive. Envy is the greatest indicator that you are discontent with something in your own life. It is the tool with which you dig up what you have buried to keep it out of sight of yourself. Don't deny the emotion, embrace it and discover what you miss. Perhaps you can do something about it, perhaps you cannot. In the former case, fix it. In the latter case, make peace with it and get on with your life. It is never about the person you envy. It is always about yourself.

Sometimes a lot harder done than said. Which is probably the reason why most people are ashamed of envy. We all feel it. Some deal with it, some let it ruin their life until bitterness dries them out

In 2005, 2009, and 2013, researchers interviewed 18,000 Australian adults. Using a scale from 1 (Does not describe me at all) to 7 (Describes me very well), the survey's participants were asked how envious they were. Almost 54 percent of respondents awarded themselves the lowest scores for envy, namely a 1 or a 2. And just over 72 percent rated themselves with a score between 1 and 3. In contrast, just over 3.6 percent scored themselves with a 6 or a 7, thereby admitting to being envious.

Such surveys are by no means proof that almost no one is envious. In fact, they are an expression of a phenomenon that social researchers refer to as "social desirability bias." When it comes to taboo topics, people are unwilling to provide honest answers, even in anonymous surveys. In such cases, pollsters need to use indirect questions to unearth people's true opinions and feelings. There is a field in psychology called scientific "envy research," and researchers agree that envy is by no means a rare phenomenon. It is widely accepted that envy has existed in all cultures and at all times, and that envy directed at successful people is extremely common.

So why are people prepared to admit other negative emotions (e.g. anger) but not envy? One reason for this is that when someone publicly admits to being motivated by envy, any actions they take to remove the cause of their envy would be deemed socially illegitimate. When envy becomes recognizable as such, or is openly communicated, then the envious person automatically disqualifies the intention of satisfying it or eliminating it. People who feel social envy never speak of envy; instead, they describe themselves as demanding "social justice." However, when they refer to "social justice," what they actually mean is "equality," which they believe can only be achieved by taking from the rich.

Envy and feelings of inferiority

The anthropologist George W. Foster asked why it is that people are able to admit to feelings of guilt, shame, pride, greed, and even anger without loss of self-esteem but it is almost impossible to admit to feelings of envy. Foster offered the following explanation: Anyone who admits to themselves and others that they are envious is also admitting that they feel inferior. It is for precisely this reason that it is so difficult to acknowledge and accept one's own envy. In recognizing envy in himself, a person is acknowledging inferiority with respect to another; he measures himself against someone else and finds himself wanting. This implied admission of inferiority, rather than the admission of envy, is so difficult for us to accept.

In citing the American psychiatrist Harry Stack Sullivan, Foster raises an issue that is of key significance in exploring the envy directed at rich people. Envy begins when one person recognizes that another person has something that they would also like to have. This necessarily leads to the question, why don't I have it? Why have they succeeded in achieving what I could not? This is a key insight. It helps us to understand why people are so vehement in denying their own feelings of envy. It also explains why most people do not want to admit that they are envious.

When they are asked, even successful people tend to explain their own achievements as the result of "luck." Nevertheless, such explanations should not be taken at face value. When people credit luck for their success, they are using a strategy designed to defend against envy. In doing so, and largely unconsciously, successful people are seeking to neutralize the envy that may be directed against them by pointing to a "random, unpredictable, and uncontrollable power that is responsible for favorable or unfavorable outcomes," or to a random combination of factors that have either favorable or unfavorable consequences for an individual.

The vile consequences of envy can be terrible in human relations and in international relations as well. I assume that one big reason for the Holocaust was envy disguised as contempt for the Jewish subhumans.

41. The sweet illusion of love resolving all our problems.

Love alone will never save the human world. The human world is cruel. Denying it indicates strongly that you are a fool. If love is the power needed in order to change humanity for the better, how come our human world gets from bad to worse? How come so many love disciples, as presented all around the world, can't make this world turn in their direction? How come these love disciples from time to another butcher other people, acting like beasts in their own lives? I find this focus on love as the elixir to a better human world infantile imagination. Rumi lived for around 800 years ago, yet his, Jesus', and other people's stress on love did not make our world better. Look at the mental suffering spreading in modern societies, the animosity between nations and people of different beliefs. Why did love not cure all these human shortcomings?

The reason for love's shortcomings is obvious if one is not deluded or an outright fool: we are very contradictory beings by our nature, and expecting one virtue as love to take over while our brains and minds keep being the same is, of course, an illusion.

42. Are we already getting dumber as to reduce our chances to find a global solution to our predicaments?

CO_2 is a byproduct of many different types of human activity and, as we continue to destroy forests and foliage that help scrub the air and provide humans and other animals with breathable air, the amount of CO_2 in our air gradually climbs. Studies have shown that too much CO_2 in the air can trigger cognitive issues, decreasing the ability of a person to focus, and hindering learning.

Getting a few breaths of oxygen-rich "fresh" air tends to clear that up, but in a future where fresh air becomes harder and harder to come by, it could lead to an overall "dumbing" of the human race. For the study, researchers simulated two different future scenarios. In both, students would perform various tasks in rooms where the air has different concentrations of CO_2. Based on established data regarding how CO_2 impacts cognition, the researchers crunched the numbers and came up with some pretty scary results.

The researchers report that in the first scenario, students were still exposed to so much CO_2 that their cognitive abilities were decreased by 25 percent by 2100. In the second, which was the business-as-usual scenario, the students were exposed to so much CO_2 when the windows were opened that they experienced a 50 percent reduction in cognitive ability.

It's just a theoretical experiment, but the scenario the researchers used assumes that humans are unable to curb their CO_2 emissions by the year 2100. We'd hope that wouldn't be the case, but if things were to play out in this way, the amount of CO_2 in the air would lead to a 50 percent reduction in cognitive ability.

A future where the human race is even dumber? It can become a fact, and it sounds scary.

The "happy end future" conclusion

"As far as I can see, we have on this planet two kinds of psychiatric wards. The first one is a place where psychiatric patients and other mentally suffering people are supposed to recuperate and regain their faculties and mental health, with not too convincing outcome. Their treatment gives mixed results, accompanied by heavy ill effects. The other kind of psychiatric ward is where most humanity exists, considering itself to be sane but acting often in an insane, irrational, destructive, and self-destructive manner. The doctors in these wards are God, angels, unbending truth-sayers, politicians, and other patented truth-endowed people trying their best but shoving us closer and closer toward the abyss without realizing it. The patients in this ward can only be cured by evolving further and away from their mental limitations" (B. K.).

What do we know about human nature that can reduce our chances of long-term survival?

Homo sapiens are basically contradictory beings, generally with lots of conflicting feelings, values, and behavioral patterns. They are creatures enslaved by their convictions while they apply their rationality best on a short-term range. On a long-term range, their rationality becomes skewed/distorted (due to its contamination by their greed, shortsightedness, and their emotions, superstitions, and convictions). We can see convincing evidence for this deficiency in the poor state of Mother Earth that we have inherited. We see it in our endless wars and conflicts, in religious and ideological conflicts/tensions/divisions, and in the lack of a wise, farsighted guiding hand in the long-term affairs of humanity.

Homo sapiens' use of cognitive dissonance, make-believe, and pretense is extensive, which is often out of their full awareness or control. *Homo sapiens* are plagued by cognitive dissonance, believing in certain values but acting often in opposition to them and their convictions. This mental process in their affairs impacts sapiens' existence on two fronts:

1) It impairs their self-view and outward judgement.
2) It harms their chances for survival as their judgment on crucial survival questions (like overpopulation and overconsumption) becomes flawed due to their cognitive dissonance.

There are exceptions among us; not so few of us can escape this trap and generate a long-term overview and sustainable lifestyle. But basically *Homo sapiens* are group thinkers. Groupthink remarkably diminishes people's capacity to exercise free, independent thinking, and thereby real free will. Also, here we find exceptions.

We, human beings, possess fine virtues, noble attributes, and we can be loving, compassionate, gentle, generous, and caring. Some of us possess genius, inventiveness, and perform groundbreaking feats, and even demonstrate a portion of long-term wisdom and aspirations far beyond the limitations of ordinary sapiens.

But our conduct reminds me too much of the coming allegory.

On a cold winter day, a little bird sat on a branch on top of a tree, singing her lovely songs with great dedication. The blowing icy wind and the falling snow made it numb, and eventually it fell down from the branch. The numb bird was fortunate to fall down into a warm shit cake a cow just left behind. The warmth from the shit cake brought life back to the little bird, and being so joyful for being saved, it started to sing again. A cat passed by, heard the beautiful song, and rushed to the bird. He ate the bird and the songs as well (Russian allegory).

So can our happy end future conclusion end in massive illusion?

Our lack of long-term attention is due to our defective comprehension, which is due to our brain's calibration, with its (for us) invisible limitations of small variations within fixed replications. This we must get away from in order to create an enduring, sustainable future for our progenies!

92

Chapter 3

THE INEVITABLE CURE TO OUR GLOBAL FOLLIES: ADVANCED, EVOLVING VISION/MISSION/ WISDOM

"In this world you can be a lifelong complainer, a tireless contemplator, a Maj fly hedonist, but also an inspirator/creator. As many humans are reality escapists, denying its essence, they become fairy tales teller and feel as both victim and omnipotent" (B. K.).

"Having too many people on earth, too much production, pollution, and consumption, the only solution is reducing them all, not relying on win-win smart fixes" (B. K.).

The main road leading to "Rome" (by Benjamin Katz)

The fundamental truths that we must face in order to survive/prevail on a long-term basis are as follows:

1. Everything is temporary, including *Homo sapiens*, his achievements, and his constructs. Sapiens is but a unit in the potential evolving process of advanced intelligent life, not the crown or the glorious end of it.

2. We are, generally speaking, unwise regarding our conduct on the planet's sustainable state. We act, in fact, as its vermin, not as its guardians.
3. Only by transcending ourselves will we pass on our valuable and noble essence to our future upgraded children. Sabotaging this transformation will imply our species' decay and oblivion.
4. Much human mental suffering is caused by modern, faster-faster lifestyle combined with human mental frailty/fragility.

The cure for the growing mental problems among people is slowing down life's tempo plus upgrading humans beyond their mental frailties.

5. The potential ultimate meaning of our existence is to further evolve as to become masters of our lives and death, and expand to far away environments in the universe and steer the process of our further evolution.

Realizing these truths, we must face the crucial question: Can we, as we are, with all our faults and shortcomings, save ourselves from our own created catastrophes? Are we capable of cooperating on a global level as to resolve our self-created global predicaments?

Look at us. With the COVID-19 pandemic and climate change, humanity is contending with global, collective threats. But for both, our response has been bogged down, less by a lack of ideas or invention than by a failure to align our actions as groups, either within nations or as a world community. We had little trouble producing effective vaccines against this scourge in record time, but how much does that matter if we can't get it to most of the world's people, and if even those who have access to the shots won't bother?

Global failures of cooperation are, of course, nothing new; we did have those two world wars. But now we're facing something perhaps even more worrying than nationalist enmity and territorial ambition. What if humanity's capacity to cooperate has been undone by our very brain's limitations and the technology we thought would bring us all together? Right now what we witness is a sour and fragmented global policy, an atmosphere of pervasive mistrust and confusion corroding institutions, and a collective retreat into the comforting bosom of confirmation bias.

We witness the rise across much of the world of conspiratorial alternate realities and intense polarization that have hampered progress on so many global problems. As a species, we are still searching for the solution that will make us cooperate on a global level for our future civilization's long-term sake. But are we able to use a solution with our divisive minds? I doubt it strongly.

What must we do instead?

We must follow Rumi's words: "Start a huge, foolish project, like Noah... It makes absolutely no difference what people think of you."

And that is exactly what we've got to do: build up a new Noah's ark for our long-term, transcending survival. In order to build such an ark (metaphorically, of course), we, or some of us, need to acquire what I dubbed as advanced, evolving wisdom.

But let us understand what wisdom is all about.

What is wisdom, and why is it badly needed in human affairs?

Wisdom is of great importance in our lives, as it helps us to deal with things in the best possible way to achieve the best results that a person seeks, taking into account all the possibilities that may arise to change the order of priorities and thus change the goal-seeking behavior.

Essential meaning of *wisdom*

1: knowledge that is gained by having many experiences in life
2: the natural ability to understand things that most other people cannot understand
3: knowledge of what is proper or reasonable; good sense or judgment

What's the difference between wisdom and knowledge? Most people think that wisdom and knowledge are the same thing but actually they are two different sides of the same coin. Knowledge is nothing but the facts known by a person, whereas wisdom is the combination of experience and knowledge with the power to apply them, or the soundness of judgement of a person.

Why does this wisdom seem to play so meager a role in our dealings with global issues such as climate change? Look at this seventy-six-year-old man. You see me, but not really me deep inside. You see a sapiens. Yes, I am sapiens, like you, but I wish to transcend this debilitating phase of my development, as I feel myself put in a mental straightjacket. Why so? Because, as all human beings, I am partly a beast, consisting of too little star dust and too much mud. I am prone to being shortsighted, self-glorifying, greedy and self-focused, pathetically self-deceptive, and a great denier of facts on account of untested convictions. I act often automatically, semiautomatically, destructively, and self-destructively, and hide these facts from my awareness through my cognitive-dissonance mental program. All these mental programs are ingrained in my brain, and self-adulation and unfounded hopes camouflage them from my scrutinizing mind. I must come out of this straightjacket before it destroys me.

Coming out of our biological, mental straightjacket is the only way we can survive/prevail in the future. We, or many of us, will need to transform ourselves in order to enhance our chances. This is what advanced, evolving wisdom is all about.

What is advanced, evolving wisdom?

It means, above all, the farsighted, sustainable, socially just, and ever-evolving thinking and praxis of future humanity. It means evolved humans beyond sapiens' limits who will open new vistas and horizons for exploring and challenging our physical and mental limitations and space.

It includes contextual thinking, rejecting fossilized stiff convictions/faith and outdated morals as a guiding star for future humanity. Its guidelines and practices will supersede all other ideologies, religions, and convictions. It includes complementary thinking, where contrasts and differences can supplement each other in creating new mental and practical order. It implies combinatory thinking, where elements for different disciplines and areas are being combined in order to create new attributes and original thinking. It includes farsightedness and the ultimate meaning to our existence: becoming sustainable and expanding creators.

It stands for recalibrating the global climate, population, consumption, and production in order to maintain global sustainability. It means strong reduction of global population, consumption, pollution, and production. We need to follow a long-term strategy for drastically reducing global population (maximum 3 billion), pollution, production, and consumption. As long as we don't reduce them, we will choke our life conditions on earth.

We badly need more farsighted, advanced, intelligent beings than the conviction-prone humans we have in abundance today (who often are slaves to wishful thinking/compulsions and greed). Therefore, it stands for upgrading many of us, who are willing and capable, into becoming emotionally/mentally stable (compared to the common mentally fragile sapiens), equipped with mindsets free of greed, shortsighted, warmongering attitudes, deception, and self-deception.

It promotes human dignity based on granting rights while demanding obligations from future beings to work for the common, long-term good of civilization. It goes for abolishing the group of the very rich/powerful, and reducing economic gaps between future human beings.

In the future, we will upgrade the followers of this wisdom to become wiser, mentally more healthy, stable, long lived, and farsighted compared to the current common sapiens. We will grant them much longer and healthier lives and meaningful engagements by dedicating themselves to pursuing and realizing our ultimate meaning: becoming creators.

Only a global order based on advanced, evolving wisdom can open great collective deeds where people can feel that they contribute to something much greater than their mundane lives. Only by joining in this huge feat of expanding our capacities and mastery of our world and destiny can we grant people the solid notion of engaging in a megalopsychia-prone life.

This advanced, evolving wisdom will not be rejected for a very long time, as it is a precondition for our future progenies' survival and further evolvement. If we won't take it as a friend, we will suffer immensely, and the survivors among us will then take it to them.

Modern technologies like CRISPR will facilitate this wisdom's promises

By starting to focus on upgrading many of us beyond our debilitating shortcomings and ailments, it will get more and more followers. Its followers will focus on how to curb physical/mental suffering in the world, and boost future humans' capacity to think long-term, farsightedly, and evolving. Suffering caused by mental disorders (like psychiatric ailments, depression, anxiety, chronic fatigue, autism, ADHD, and other ailments) will be reduced as to boost people's capacity to think/act beyond their own spheres.

This wisdom is not dictatorial, but it will weaken deceptive notions of baseless personal uniqueness so prevalent among ordinary people in our mass societies, and presents instead down-to-earth engagements on how to attain megalopsychia.

Being practiced properly, it will markedly reduce human folly, mediocrity, and their negative omnipotent manifestations in our lives

As the most fundamental issue for us is our long-term evolution, sustainable survival, which is dependent on many of us attaining advanced, evolving wisdom, we hope that in due time (several centuries) we will be able to save ourselves from our own follies/vice.

What is the ultimate meaning of our lives from the viewpoint of advanced, evolving wisdom?

"To be in the world but not of the world, and thereby create new, ever-evolving paths to other worlds is the greatest meaning granted advanced intelligent life" (B. K.).

Jalāl al-Dīn Rūmī, an acclaimed Sufi mystic and poet, carved this beautiful and insightful sentence: "I died as inert matter and became a plant. And as a plant I died and became an animal. I died as an animal, and became a man. So why should I fear losing my 'human' character? I shall die as a man, to rise in angelic form." Jalāl al-Dīn Rūmī saw human purpose as, essentially, evolutionary. Ergo, the ultimate meaning of our lives must be further and constant evolvement beyond our own limits as to make us match as well as we can one day our image of our Creator. This is our ultimate meaning!

There is an evolving potentiality in nature/the universe indicating how the journey of farsighted, intelligent life should be conducted, and how intelligent life should emancipate itself from the bonds of life's dust, temporariness, and limitations. This evolving potentiality does not back up the false omnipotent view that common *Homo sapiens* nurture regarding their selves as being irreplaceable or indispensable in the world and in the process of further evolvement.

This evolving potentiality is a dormant force that opens for us—or some of us—the possibility to pursue an evolving journey beyond God and *Homo sapiens*, all the way to the stars, in order to become creators by our own merit. The ultimate meaning of our lives is, thus, to join in this journey and steer it as well as we can, constantly transforming and transcending ourselves and our expanding worlds, yet doing so in a sustainable manner.

The message—of what advanced life is all about—had been encapsulated in us, waiting for its carriers (us) to mature and to gain awareness beyond our current mindsets and our fantasy-prone constructs so it could reach its own fruition in our minds and deeds. The animated dust strives after becoming the creator. Yet this potentiality is not coded as an imperative but as a potential, as the universe is basically indifferent to our ultimate meaning.

How can we maximize our chances of long-term survival?

1) Having too many people on earth (by now, 2022, we are 8 billion) we must get down to less than 4 billion within 400 years if we wish to survive without becoming dumb and sick.
2) We must drastically reduce global consumption, production, and pollution.
3) We must stop the mass depletion of Earth's life-sustaining resources.

4) We must practice strict policies aimed at reducing/leveling out the economic/social gaps among citizens, and by binding humans rights (decent existence) to human obligations for the long-term common good.

5) We will improve humans' lives, mental and physical health, and wisdom through genetic engineering and editing.

6) We will introduce artificial intelligence to help us upgrade sapiens to the stage of creators.

7) All these goals will be achieved by slowly establishing a global governing organ empowered by many entities, with real political power and military force to enact these policies.

What are the essentials of my vision?

"Without the aspiration to attain greatness and further evolvement, sapiens is but a dead end" (B. K.).

The essentials of my vision/mission implies a global, sustainable, trimmed-down civilization of a maximum of three to four billion citizens within a maximum timespan of four to five hundred years, being steered by a political system called global democrature. *Democrature* means global dictatorship with military might, dealing with all critical global issues regarding sustainability and evolution, while the regional/local issues are dealt by different political systems of people's choice similar to democracy.

Regarding a plan to reduce global populations, it will be forced upon individuals, societies, and humanity in the form of an agreed quota bound to the ecological footprint per capita. It will be done by carrot-and-stick method. Regardless of the measures that will be introduced to spare people from mass starvations and climatic catastrophes, countless lives will be lost in the interim period due the consequences of climate change, to natural disasters, wars, epidemics, hunger, and lack of drinking water. All these plagues will surely hit our civilization within the next 200 years. The global democrature (global governance) is a must in order to achieve our upgrading and to pass as painlessly as possible the critical phase of stopping the global climate from going awry.

Humanity will have to go through a difficult period of hundreds of years of bloodshed, tears, suffering, and sacrifices. We must expect that in the establishing and cementing phases of this new civilization, many negative manifestations of atrocious human behavior will surface in order to sabotage/derail this project and entertain narrow self-interest. They will have to be subdued by all means. A formidable combat force will assist in implementing needed decisions in case an entity tries to work against the principles of this new vision. Democracy, as we know it today from the west, will only serve as a political system for regional/local communities. This new democracy will focus on the enlightenment of its citizens, farsighted perspectives, and pursuing the survival strategies and goals of the vision. In the phases of building a new civilization, artificial intelligence will become so advanced as to surpass humans' mental capabilities in certain areas—like long-term projections, analysis, complex interacting systems (including weather patterns)—and therefore will be integrated in the functioning of the global governance and controlled by it as well.

This vision is bound to our farsighted global vision for the ultimate meaning of our lives. It is bound to global sustainability, our further evolvement, and values like prudence, modesty, and fairness in all life aspects.

A model for what a future this global vision based on advanced, evolving wisdom implies is depicted below.

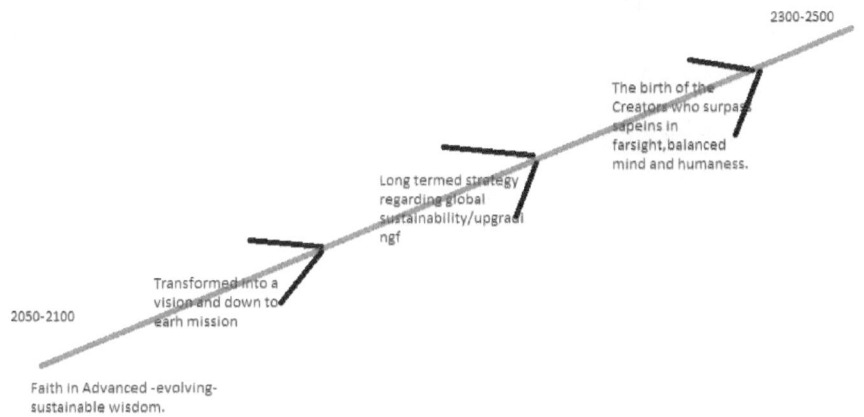

Quotas and ecological footprints in the future

In the period of recalibrating the global climate, population, production, consumption, pollution, and civilization's core values, austerity measures will stretch from birth control (how many children a fertile person is entitled to bring to the world in the future) to ecological (consumption/production, pollution) footprint/quota. If a person exceeds this quota, he will be fined or punished more severely (imprisoned) in accordance with the severity of his crime.

But who and how will these measures be enforced? Who enforces major mobilization in case of a war? Who enforces taxes on people, sends people to prison for their crime, children to school, and the like? You know the answer: enforcing organs such as governments, with laws on their side. This will also be the case then in the future.

Will it be difficult to implement? Probably, as we will end up becoming in this millennium both upgraded, advanced beings enhanced by cyborg technology (https://medicalfuturist.com/superhumans-2021/). As a matter of fact, enhancing human capabilities has been on the minds of people for ages, but it came a long way from ancient training methods to exoskeletons. Enhancing our abilities, be it permanently or temporarily, is a tempting but risky matter. For will it be a possible reality in the not-too-far future that we need to gear up superhuman powers/brains so that we do not lag behind the others? Let's dive in and see how (and if) we can be our better self.

The dawn of digital/genetic health brought an era of technological advancements, and the recreation of the human body or body parts or functions came within reach. It started as a restoration of a certain human function, like eyesight or hearing, or the replacement of a missing limb. Technologies like CRISPR gene editing, nanotechnology, and even the evolution of wearables and sensors brought new possibilities in human enhancement. And although a 200-year longevity is still far away for many, living longer and possibly healthier is a de facto case for the human race. (That is, if we survive the climate catastrophe we've created.) We ought to change human life spans from flailing, promising youthful vigor followed by physical/mental decline into a constructive, evolving, and enduring life span shining throughout. This will be both a difficult and possible task!

Can we reprogram existing life at will?

The answer is yes. The central code for biology is simple. DNA letters, in groups of three, are translated into amino acids—Lego blocks that make proteins. Proteins build our bodies, regulate our metabolism, and allow us to function as living beings. Designing custom proteins often means you can redesign small aspects of life. For example, getting bacteria to pump out life-saving drugs like insulin, or a developing a human to become much wiser and farsighted.

All life on Earth follows this rule: a combination of sixty-four DNA triplet codes, or "codons," are translated into twenty amino acids. Why wouldn't sixty-four dedicated codons make sixty-four amino acids? The reason is redundancy. Life evolved so that multiple codons often make the same amino acid. So what if we tap into those redundant, "extra" codons of all living beings and instead insert our own code?

A team at the University of Cambridge recently did just that. In a technological *tour de force*, they used CRISPR to replace over 18,000 codons with synthetic amino acids that don't exist anywhere in the natural world. The result is a bacteria that's virtually resistant to all viral infections because it lacks the normal protein "door handles" that viruses need to infect the cell. But that's just the beginning of engineering life's superpowers. Until now, scientists have only been able to slip one designer amino acid into a living organism. The new work opens the door to hacking multiple existing codons at once, copyediting at least three synthetic amino acids at the same time. And when it's three out of twenty. That's enough to fundamentally rewrite life as it exists on Earth.

Hacking the DNA code

Our genetic code underlies life, inheritance, and evolution. But it only works with the help of proteins. The program for translating genes, written in DNA's four letters, into the actual building blocks of life relies on a full cellular decryption factory.

It's a great idea in theory, but a truly daunting task in practice. It means that the team has to go into a cell and replace every single codon they want to reprogram. A few years back, the same group showed that it's possible in *E. Coli*, the lab and pharmaceutical's favorite bug. At that time, the team made an astronomical leap in synthetic biology by synthesizing the entire *E. Coli* genome from scratch. During the process, they also played around with the natural genome, simplifying it by replacing some amino acid codons with their synonyms—say, removing TCGs and replacing them with AGCs. Even with the modifications, the bacteria were able to thrive and reproduce easily. The superpowered strain, Syn61.Δ3(ev5), is basically a bacterial X-Men that grows rapidly and is resistant to a cocktail of different viruses that normally infect bacteria.

Because all of biology uses the same genetic code, the same sixty-four codons and the same twenty amino acids, that means viruses also use the same code. They use the cell's machinery to build the viral proteins to reproduce the virus. Now that the bacteria cell can no longer read nature's standard genetic code, the virus can no longer tap into the bacterial machinery to reproduce, meaning the engineered cells are now resistant to being hijacked by almost any viral invader.

These bacteria may be turned into renewable and programmable factories that produce a wide range of new molecules with novel properties, which could have benefits for biotechnology and medicine, including making new drugs, such as new antibiotics. Viral infection aside, the study rewrites what's possible for synthetic biology. Perhaps the most exciting prospect is the ability to dramatically rewrite existing life. Similar to bacteria, we—and all life in the biosphere—operate on the same biological code. The study now shows it's possible to get past the hurdle of only twenty amino acids making up the building blocks of life by tapping into our natural biological processes.

Next up, the team is looking to potentially further reprogram our natural biological code to encode even more synthetic protein building blocks into bacterial cells. They'll also move toward other cells—mammalian, for example—to see if it's possible to compress our genetic code(14).

By using CRISPR today, we may soon eliminate malaria (2022). By using CRISPR tomorrow (from around 2040–2050) we may reduce/eliminate human stupidity, which like malaria is a deadly scourge.

Humans have genetically engineered animals, corn, wheat, rice, fruits, and vegetables but are afraid of doing the same with themselves, even when the gains outweigh greatly the drawbacks (think of the immense gains for our evolving civilization in eradicating/reducing human stupidity in personal/social, national/ global affairs).

Brainpower

A microchip in the brain that will enhance our cognitive functions by a zillion—how we all could use such a thing during exams! The concept was first raised back in 1968 by architect Nicholas Negroponte at MIT, and it has come a long way since. Hacking the human brain is more difficult than we could imagine, and cannot be solved with a single microchip implant.

Brain-computer interface

Researchers have been investigating the possibilities of brain-computer interfaces, or BCIs, for quite a while now. At first, these were thought to be tools on one hand to provide constant monitoring of the brain's electrical activity; this could support a wide range of applications from monitoring epilepsy or ADHD to pain management and sleep assessment. On the other hand, the concept was thought to be a solution for paralyzed people to move and control things around them with only thoughts. Taking it to the next level, such an implant could also be used as an external hard drive for the mind.

Enhancing cognitive performance would be able to change the lives of millions suffering from memory loss, neurological or psychological disorders. Restorative processes like stroke rehabilitation would also gain from the technology. Or, as a matter of fact, these can help reduce the cognitive effects of ageing. No wonder there are many studies that focus on the possibilities such technology can provide for seniors.

3D printers are already printing prototypes of near-perfect internal organs, genetic engineering will grow alternative organs for us, genetic research will eliminate quite a few terrible diseases—all of which are already in various stages of development today. But there will be quite a few more issues such as bionic organs, remote diagnosis, artificial intelligence that will scan huge databases in the blink of an eye, and more.

It will be difficult, but I am sure that as it is being done nowadays, it will become globally sanctioned. The duality of human existence is always an issue; what good/evil intent or unintentional consequences it may portend. But the evolving direction toward becoming a creator constitutes our ultimate meaning.

Long-term life-affirming pragmatism points strongly toward sustainability, our further evolvement, and the upgrading of our brains as to become much wiser than we are today—creatures free of our recurrent wars, excesses, conflicts, and divisions.

The science of it also has a philosophical backing called transhumanism, a movement that "advocates for the transformation of the human condition by developing and making widely available sophisticated technologies able to greatly modify or enhance human intellect and physiology." The concept states that humankind will evolve into a new intelligent species.

How can advanced, evolving wisdom save us from ourselves?

There are interesting observations and references regarding AI and different kinds of human intelligence. But how we are to make this existing phenomena—advanced, evolving wisdom—operational/useful in saving us from our own vices is the most relevant question for us in a menacing period of alarming climate change, pollution, overconsumption, and overpopulation.

With respect to our long-term survival, I find that advanced, evolving wisdom to be connected with farsightedness, allowing us to see different contingencies of the potential future of humanity in ever-changing contexts, and derive out of it a strategy or grand plan as to how to steer our further evolvement in a sustainable manner. This is our cardinal challenge, and we need to focus on such wisdom that can remove us away from our own created abyss.

Chapter 4

CAN WE FACILITATE THIS NEW VISION/MISSION/WISDOM?

"We cannot change our self-destructive course with our old or new invented stories/religions/ideologies unless we evolve beyond our limitations and combine it with teaching people how to think right. Educating people right is to teach them how to think independently, critically, sustainably, evolvingly, and beyond their mental/conceptual box" (B. K.).

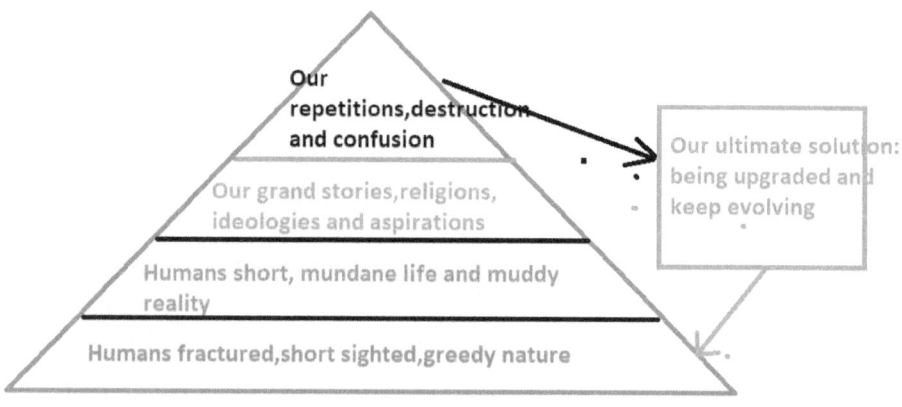

A dynamic model for how to change human
nature for the better (Benjamin Katz)

Look at nature, at its complexities, contradictions, cruelty, remedies beauty, and evolving force. By realizing its complex essence, you get much of the sense of this world. Then you shift your focus to the greatest gift of all that we have been granted: the capacity to think out of nature's box, the capacity to extend it, to create and devise new things in it, and to modify by our own free will/ ingenuity both this world and ourselves. When you grasp this overview, you may perceive both the blinding impact of our mental straightjacket and the immense possibilities open to us if we transcend sapiens.

The tale of the sand is one of the more subtle, sublime, and gracious stories telling us about what we can really become by transcending our sapiens' limitations.

A stream flowed from its source in the mountains to reach, at last, the sands of the desert. Just as it had crossed every other barrier, the stream tried to cross this one, but its waters disappeared, soaking into the sand. A hidden voice, coming from the desert itself, whispered to it that the wind that crosses the desert can do this work for it. The stream objected; the wind could fly, and this was why it could cross a desert. The voice told the stream that by hurtling in its accustomed way it would not be able to cross the desert. It had to allow the wind to carry it over to its destination. But the stream did not understand how this could happen. The voice told it that by allowing itself to be absorbed by it could accomplish the journey. The stream could not accept the idea as it had never before tried it. It was afraid to lose its identity. The voice told the stream that if it did not believe in this possibility, it could not become more than a quagmire, and even that could take many, many years. The stream wanted to stay the same stream it was today. The voice told it that it could, in either case, remain as it was, but its essential part would be carried away and formed into a stream again. When the stream heard this, it could dimly remember a state in which it had been held in the arms of wind. It also remembered that this was the right thing to do, though not necessarily the obvious thing to do. And the stream raised his vapor into the wind, which bore it upward and along, letting it fall softly as soon as they reached the mountains. And because the stream had had its doubts, it was able to remember the details of the experience. He reflected over this experience, concluding, "Now I have learned my true identity."

This is the essence of our transcending, but it will be full of struggles, wars, blood, and sufferings because sapiens will not willingly give up their dominance on the planet Earth, even while destroying their life condition and themselves.

The only way to touch eternity is for us to keep on ever evolving in order to attain creators stage, thereby dissolving our biological limitations

This is the way, a and transcending ourselves via the tools of advanced, evolving wisdom is the pillar of my vision!

"If man will not hitch his wagon to the stars, he will die traceless behind sapiens' prison bars. My vision will be first accepted when panic/desperation in global affairs overshadows the false hopes for a win-win solution" (B. K.).

The forecast of how much time we have left in our technological future could then follow from statistical information about the fate of civilizations like ours that predated us and lived under similar physical constraints. Most stars formed billions of years before the sun, and may have fostered technological civilizations on their habitable planets that have perished by now. If we had historical data on the life span of a large number of them, we could have calculated the likelihood of our civilization's survival in different periods of time. The approach would be similar to calibrating the likelihood of a radioactive atom to decay based on the documented behavior of numerous other atoms of the same type. In principle, we could gather related data by engaging in space archaeology and searching the sky for relics of dead technological civilizations. This would presume that the fate of our civilization is dictated by the physical constraints.

But once confronted with the probability distribution for survival, the human spirit may choose to defy all odds and behave as a statistical outlier. For example, our chance of survival could improve if some people choose to move away from Earth. Currently, all our eggs are in one basket. Venturing into space offers the advantage of preserving our civilization from a single-planet disaster. Although Earth serves as a comfortable home at the moment, we will ultimately be forced to relocate because the sun will boil off all liquid water on our planet's surface within a billion years. Establishing multiple communities of humans on other worlds would resemble the duplication of the Bible by the Gutenberg printing press around 1455, which prevented loss of precious content through a single-point catastrophe.

Of course, even a short-distance travel from Earth to Mars raises major health hazards from cosmic rays, energetic solar particles, UV radiation, lack of a breathable atmosphere, and low gravity. Overcoming the challenges of settling on Mars will also improve our ability to recognize terraformed planets around other stars based on our own experience. One can argue that we have enough problems at home and ask: "Why waste valuable time and money on space ventures that are not devoted to our most urgent needs right here on planet Earth?"

Before surrendering to this premise, we should recognize that attending strictly to mundane goals will not provide us with the broader skill set necessary to adapt to changing circumstances in the long run. A narrow focus on temporary irritants would resemble historical obsessions that ended up being irrelevant, such as "How can we remove the increasing volumes of horse manure from city streets?" before the automobile was invented, or "How do you construct a huge physical grid of telephone landlines?" before the cell phone was invented.

True, we must focus our immediate attention on local problems, but we also need inspiration that elevates our perspective to a grander scale and opens new horizons. Narrowing our field of view drives us to conflicts because it amplifies our differences and limited resources. Instead, a broader perspective fosters cooperation in response to global challenges. And there is no better fit for such a perspective than science, the "infinite-sum game" that can extend the life span of humanity. As Oscar Wilde noted: "We are all in the gutter, but some of us are looking at the stars." Here's hoping for more of us looking up. The inspiration gained from that view might carry us well beyond the statistical forecast that impels the fatalistic alternative.

But while speculating on all these dim possibilities, we must teach people, in any conceivable manner, including upgrading their mental faculties, how to learn and focus on the essentials of them prevailing and surviving the menacing future, which they created on earth.

In the story "The Making of Fire," the metaphor is self-evident: How do you learn to facilitate essential and unifying knowledge for different groups of people differing in faiths, convictions, rituals, and contexts.

This knowledge of teaching how to make fire (metaphorically) in shifting contexts is available to many of us. The capacity to teach it to the many who resist it, encapsulating themselves in their own special dogmas, is our great challenge.

Chapter 5

First lessons in the University of Attaining Advanced, Evolving Wisdom

The meaning of human life is bound to impending global necessity; to prevail/survive/thrive, ergo to evolve into farsighted and wise beings, which we, sapiens, are not. Troubleshooting essential global matters must include the minds of the shooters, as they seem to aim at the effects, not at the cause!

As I pointed out before in the book, advanced, evolving wisdom consists of the capacity to think proactively also as a warrior struggling for a future for advanced, intelligent beings beyond sapiens. To think contextually, complementary, combinatoric, "out of the box" creatively, farsightedly, evolvingly, and also by discerning between essential issues for our long-term survival on the one hand, and secondary and nonessential issues on the other hand.

Below I present you with small texts in sort of chaotic order (to prevent you from categorizing in the common sapiens' manner) aimed at training your mind to comprehend this wisdom's scopes and vistas, be it self-knowledge, knowledge of the world/reality, knowledge of our future journey/ultimate meaning, and their inevitable intertwining.

Revelation (photograph by Benjamin Katz)

Let us start with small comments on our lack of clear meaning with our life as species, our fragmented and conflicting nature, and thereby our chaotic global state.

If not evolving away from our crude human/beast, what is our purpose without wisdom and peace?

The core problem of humanity is not religions/ideologies but human nature, which is shortsighted/fractured and contradictory. We cannot change our self-destructive course with our old or new invented stories unless we evolve beyond our limitations and combine it with teaching them how to think/behave rightly.

Educating people rightly is to teach them to think independently, critically, sustainably, evolvingly, and beyond their mental/conceptual box.

Can we recalibrate global climate?

When will we be free of humanity's infernal noise, its lack of poise, its misguided voice, which our world destroys?

Before you assess our chances to recalibrate the climate, remember that we talk much better than we walk and we are self-deceivers. Remember that Illusions, delusions, wishful thinking, and self-deception constitute the foundation of our faulty and misleading mind/brain. Remember that "I believe in miracles 'cause it is absurd" is the essence of human (not necessarily facts based) convictions, be them in gods, reincarnation, or in science/technology. Remember that soothing lies taste better than bitter truth for the unscrutinizing, infantile mind. It's easier for people in general to believe a soothing lie than to have one's mind changed by information that is new and novel. This means that if you believe something for political or religious reasons, it's far harder to change a person's mind and have them understand a fact that differs from that person's opinion.

Remember that humans always take sides, more often than not the side of escapism, hedonism, easy short-term solutions than the struggle against external and internal seductive oppressors, resulting in them being victims or generating victims.

Remember that we create and generate frictions all the time.

Splitting up on views, values, interests, convictions, and faith is sapiens' mental Achilles' heel.

Whether there is one God or more or none at all is secondary compared to our incapacity to agree on the essentials in our lives.

The core problem of humanity is not religions/ideologies but human nature, which is shortsighted/fractured and contradictory. People are limited beings. Can they get past selfishness, ignorance, animus, and worst of all, evil without transcending themselves? Without evolving beyond sapiens' constraints, humanity is likely to follow Shakespeare's sentence, "Out, out brief candle."

If there was a global psychiatric ward for nations due to crazy/conflict filled conduct, there will be more nations in it than outside it.

Remember that educating people rightly is to teach them to think independently, critically, sustainably in all aspects, and beyond their mental/conceptual box.

In the face of mortal danger like climate change, many follow the course of passivity, encouraging thereby our self-destructive behavior.

Therefore COP 26 commitments to stop climate change will, at best, amount to under 50 percent of proper action due to *mundus vult decipi*, faulty strategies, wrong priorities and execution, and bad global cooperation. As long as there are rivalries/tensions and arms' race between nations, global climate will deteriorate. Self-interest is the core of the deteriorating climate change.

There are five holy cows that we must slaughter in order to slowly remove the mortal danger of climate change: over population/consumption/production, free market practice, and our delusional mindsets that make many of us believe that we are the end of intelligent evolution.

Yet poor sapiens keep mentally wiggling and wriggling to avoid the inevitable for his own rescue: cutting down on population, consumption, production, and pollution.

Let us continue with some basic guidelines:

1) You shall abstain from damaging others, your life conditions, and yourself, and stop bearing a pretentious mask. Wearing your pretentious social mask is a burdensome task.

Who are you beyond this mask? If you have not killed, harmed people, or damaged life conditions at will, and are not abusive and self-abusive, you have managed the first stage of attaining advanced, evolving wisdom. You can't spring over this stage.

2) You have to attain some mental balance. Being plagued by mental ups and downs severely affects your overview, outlook, and capacity to learn and absorb teaching beyond your own self. Achieving a healthy mind in a healthy (not drug abused) body grants you a chance at attaining advanced, evolving wisdom.

3) Beware man: You are not exceptional/unique as an individual/nation unless you do something enduringly exceptional for humanity's long-term evolution's sake. Not contributing in this respect and just living your own life makes you, at the most, an ordinary human being. If you wish to know how significant you are in this stage of ordinary life, put your finger into a glass with water, then withdraw it. The hole which is left tells how important you are.

4) The perfectionist trap: If you refine too much, you conclude nothing. Realizing great endeavors and projects cannot be elegant, clean, and always win-win may help you to act with calculated risk. Yet, short-term gain may implicate long-term pain, and you must learn to discern between the two.

5) How does unsustainable population affect climate change?

More people means more demand for oil, gas, coal, and other fuels mined or drilled from below the Earth's surface that, when burned, spew enough carbon dioxide (CO_2) into the atmosphere to trap warm air inside like a greenhouse.

During that time, emissions of CO_2, the leading greenhouse gas, grows twelvefold (July 29, 2009). Besides, more people = more use of all minerals, metals, concrete, sewage, water, soil, ergo more pollution and burden on natural resources. Population and climate change are inextricably linked. Every additional person increases carbon emissions—the rich far more than the poor—and increases the number of climate change victims—the poor far more than the rich. Cutting down our overpopulation while fixing a quota per capita of greenhouse gas emission and the use of polluting methods and resources will be the optimal solution to reversing climate change over generations.

6) How should global overpopulation be dealt with?

Having around 1.9 billion people on earth will ensure a decent life standard. If Australians want to continue living as we do without making any changes, and as a planet we want to meet our footprint, then the number of humans Earth can sustain long-term is around 1.9 billion people, which was roughly the global population 100 years ago in 1919 (July 25, 2019).

At the moment, 2022, there are 8 billion people living on earth, a disastrous rise by 5.5 billion since 1950. Yet no serious politicians consider the idea of reducing global population using the stick and carrot method, even though this will be the most enduring, wise solution to the problems of pollution, food supply, clean water, dwindling resources, desertification of huge areas, fatal climate change, and the dangerous rise in temperature.

First and foremost, we have to undermine resistance for this idea by its powerful, anachronistic lobbyists: the religious lobbyists and human rights lobbyists. The reason reducing population to combat climate change is nonexistent is the religious/human rights lobbyists acting like the cigarette/oil industry and firearm lobbyists, which used billions of dollars to derail any discussion of their crimes. We don't know how much our convictions dumbs us down where we try to save ourselves from our self-created climate change.

It can be done by a global organ carrying out a policy of stick/carrot regarding giving birth. The organ will be equipped with legal legitimation and military force to enforce this policy, where people wishing to have children will have a fixed quota, say for a couple maximum of three children and for a single parent one child, and if they will exceed this, they will be punished severely with the removal of the child, a big fine/prison, and castration. Whereas if they keep to their quota, they will be rewarded.

Mental exercises in enhancing your self-knowledge, knowledge of your world and that of the ultimate meaning of our life

It is hard for the majority of people to imagine, that most of us are locked up in our contemporary mental/cognitive/brain capacity matrix and habitual behavior. The possibility of upgraded humans who will become much wiser than us is both blasphemous and undesirable for them. If you asked people 200 years ago whether it was possible to fly or to move by train, or to double our average life expectancy, they would reject these possibilities as well. What can we learn from this observation? That it proves that mental limitations define the mental "straightjacket" of man.

Compulsions, obsessions, all kinds of abuse/excesses, factually baseless convictions, and automatic rituals equal the debilitating mental straightjacket of man.

Self-knowledge:

Every day, each and every day
I'll escape my destiny; a fly of May.
I plan great deeds and to myself say:
Tomorrow these deeds, realize I may.
But then I dream of a sunny bay
where I all day long seduce and play,
where temptations my heart do sway
and so goes my willpower astray.
Although I plead: Oh with me stay!
Although for strength to God I pray
I wake up to a new day as I am: a fly of May.

Knowledge of the world:

Eternity is temporary and so is sapiens. Change and evolution may be enduring.

Is it true today and has always been the case that humanity revolves around superfluous pursuits, mediocrity among the masses, and blind ignorance? It's what the smart ones feed off of.

Self-knowledge and knowledge of the world:

Mitigating climate change is only slowing its ill effects. Only reducing population, production, consumption will do the trick. In order to do just this, we need much wiser, farsighted, and mentally balanced human beings than the current sapiens are.

Unfortunately, lots of westerners, including Danes, view other people abroad as heroes when they go against their regime's policies, but they never get out of their cozy chairs to do the same in their countries. In this century alone, USA and NATO have waged savage wars in Iraq, Libya, Yemen, Afghanistan, and also in Syria, killing and maiming millions, but no heroes rose here to protest against these war crimes. There is so much to challenge in the way we deal with things in the peaceful west, but what do we do? We talk about other nations we know nothing of, and often start stupid wars there, butchering innocent people and soldiers alike. And yet we tell them what is the right thing to do. This is the hypocrisy of the western man.

As part of NATO expeditions in the last twenty-one years of this century, we in the west have killed 1,000 times more people, innocents and children, than Israel did in all its seventy-five years of existence. NATO's current policy, retreating from Afghanistan so stupidly, resulted now (October 2021) in immense misery and the danger of hunger for 30 million people there. Who is the real criminal? We in the west are the perfumed criminals starting war after war, abusing world resources for a very long time, and contributing greatly to the global climate crisis caused by our lifestyle. Are we the hypocrites?

I do thank the thieves for stealing, as without them we would not have a police force, and without a police force all humans would become thieves.

Convictions not based on hard facts reflect both sapiens' mental frailty and force at the same time. Can you figure out why?

Wise and foolish rules of conduct

Acceptance:

Acceptance of different views and people can be granted when they don't harm people, life conditions, and our future prospects.

"It is not the man who has too little that is poor, but the one who hankers after more" (Seneca).

Envy:

There are lots of repressed people due to the norms, values, consensus that they have acquired. They are often envious of the people who feel free to unfold themselves and their talents. But as envy is considered an ugly attitude, they become experts at criticizing them for the most petty reasons.

Repressed people are easily threatened by those who throw off society's shackles or accomplish something they can't, as they themselves have either been beaten down or lack the courage to not conform. They are trapped in their own minds.

Spirituality:

The key to my agenda is that spirituality is a higher attribute than believing in something between heaven and earth as a convention/compulsion being adapted by the herd. Spirituality is striving to develop beyond the shackles of our mind/body and the current conventions also regarding our current spirituality.

This spirituality will be widely perceived when we are of humanity's infernal noise, its lack of poise, its misguided voice, which our world destroys.

Human spirituality, religiosity, and intellectualism are but collective masturbation, cutting off the driving force of our further evolution.

Compulsion:

The French have an expression, *l'appel du vide* (the call of the void), to refer to the compulsive urge in humans to do something self-destructive, such as leaping off a cliff. Often, but not always, it is satisfying just to contemplate the act yet resisting it.

Another human urge is overdoing things, often for a different reason, like living a healthy life and being granted a long life. In my country, lots of Danish people have become obsessed with physical exercises and the right intake of food, causing themselves physical and mental damage. These two urges, being so different in their focus, have a common mental core: compulsion!

Many people become foodies, self-pleasure or just casual hobbies focused to an extreme level, and thus become insignificant beings guided by false freedom and self-focus/lack of higher motives

There are these new humans working for the IT temple, celebrating their freedom, small pleasures, avoiding any emotional obligations. They are too busy and ego/pocketbook rewarded to overlook or avoid the other sides of life.

In the future:

In the future, people must be rewarded for their long-standing service/ personal commitment to a sustainable, fair, and just world

Self-knowledge:

The best deal one can strive for in life is to have proactive ideals, a mission/ vision to work for and pursue, and a reality that challenges you to work together with others toward higher goals than your own petty life, and engage yourself in it as to better it by unfolding your best potential.

Wise and foolish rules of conduct:

One of the most potent and illusory rules of human ego is the conviction that they are better than others.

Knowledge of the world:

"Having too many people on earth, too much production, pollution, and consumption, the only solution is reducing them all, not relying on win-win illusory smart fixes" (B. K.).

Self-knowledge:

Billions of *Homo sapiens* have sought refuge in the vague idea of some god/so-called truth-proven ideologies, and thereby caused irreparable damage throughout human history.

Self-knowledge and knowledge of the world:

If you expect love, compassion, and harmony to reign one day in sapiens' world with our current brains/minds, you are delusional about what sapiens are.

Much of our efforts at beautifying our reality follow the principle *mundus vult decipi* (the world wants to be deceived) in all its different manifestations, including the emperor's new garments.

Self-knowledge:

Enjoying life as an ultimate purpose in itself is infantile. I'd take care of my physical safety, mental balance, and good health while dealing with *Tikkun Olam/ Adam* (the improvement of man and our world). This is what keeps us on track toward megalopsychia, our ultimate meaning.

The more ritualized/compulsive a human becomes, the more he distances himself from knowing himself and his world

Pursuing our ultimate meaning:

Oh, blinded sapiens!
In Paradise on earth you believe
as yourself/others you deceive,
while your meaning and the others'
is in evolving further-further.

"And yet it moves" or "Albeit it does move" is a phrase attributed to the Italian mathematician, physicist, and philosopher Galileo Galilei (1564–1642) in 1633 after being forced to recant his claims that the Earth moves around the sun, rather than the opposite. My motto is similar, though with a different focus: that without becoming a sustainable and a much wiser civilization in this and the next century, we are doomed. And therefore our only chance at long-term survival is to upgrade ourselves and evolve further beyond sapiens.

Choosing between self- destructive doom and a long term survival we will choose my solution, I am certain!

Humans have genetically engineered animals, corn, wheat, rice, fruits, and vegetables but are afraid to do the same to themselves, even when the gains clearly outweigh the drawbacks.

Yet, I am sure it is being done and may become government sanctioned, depending on what it specifically entails. The duality of human existence is always an issue. What good/evil intent or unintentional consequences may it portend?

Self-knowledge:

People pursuing their impulses without a sense of mutuality, compassion, and responsibility are mass produced nowadays. People who meet a stranger and after a very short while ask for sex are fucked up in their heads and approach to close relations. Due to impulse-prone individualism, availability of pornography/drugs, casual sex, and promiscuity, lots of people in our time are psychologically screwed up. They are not able to create and maintain enduring and mutual relations, as they are disturbed/infantilized in this very essential area.

Knowledge of the world:

Self-deception is man's unrecognized socially induced mental schemata originating from illusions, delusions, untested convictions, boasting, and self- glorification.

This has led to collective stupidity, which is the most dangerous threat to us as a species in our time, as it results in both intended and unintended immense evil.

Know your world:

Those of us who have acquired advanced knowledge in mass psychology, including mass brainwashing and manipulation, probably know that these mechanisms—which we, humans, are very susceptible to due to the fact that we are social animals—are used all around the world, including in our so-called enlightened democracies. They should glue and coerce individuals together around a common narrative and consensus. Religion does the trick by the repetition of prayers, by replication of so-called truths. Our democratic institutions, of course, massively use these mechanisms of repetition and replication in our day to day in order to fortify our convictions that we are the best, democracy is the political form of the future, and we are so advanced as to choose our leaders, discard them, and hence have found the best political system on earth. As with prayers, it may take longer than a lifetime to get evidence to support these claims, but in the meantime, the chickens get hypnotized!

Brainwashed economic thinking:

Our view of economic growth by becoming more effective is also a kind of mumbo-jumbo talk promoted by our economists.

They claimed for a long time that increase in efficiency—like burning coal more efficiently—was a good thing. But as the efficiency of burning coal grew, so did the demand for more and more coal, and far beyond the initial quantity used before using more efficient measures. This paradox is visible in the history of technological improvements of all kinds. Better cars, miles per gallon, more miles driven. Faster computer times, more time spent on the computer. Cheaper energy to fly passenger flights, more flights and more passengers. Thus, the idea that efficiency is always good is redundant. Efficiency has to be put against the health of our planet, biosphere, and humans/animals all bound together.

Pursuing the ultimate meaning:

Being entirely down-to-earth, concrete blocks for the potentiality of becoming greater, wiser, and evolving without a vision that lifts our lives beyond its current phase, your life lacks a weighty mission

Shrinking one's universe to a comfortable size means become partially blinded and often useless for the long-term common good.

Great hopes for both men and humanity cannot be realized without prolonged struggle, implicating blood, tears, suffering, and sacrifices.

In the battle to gain constructive self-knowledge/self-determination and further evolvement, our brains, with their current capacity and mental programs/ habits, can be our Trojan horse.

Knowledge of the world:

History has taught those who think well that popular, seductive ideas regarding our future, without exerting protracted efforts and sacrifices, will be found useless. Effortless win-win is a Disneyland illusion.

It is obvious that the talkative chatter that characterizes modern societies is, to a high degree, bullshit and replication.

The problem in the EU and Western Europe is too much talk, too little walk. It needs armed forces of its own in order to be free of USA's hegemony. But by not acting upon this, it is a kind of vassal.

Newspapers/mass media focus on hair splitting actuality on account of strategic thinking regarding our direction as evolving species.

Self-knowledge/knowledge of the world:

Without great knowledge of human nature and its pitfalls, extensive knowledge of history, evolution, political reality, and nations' obsession with self-interest, you are prone to come up with useless, wishful ideas on how to resolve the global climate crisis.

"The problem with the world is that the intelligent people are full of doubts, while the stupid ones are full of confidence" —Charles Bukowski

He, Charles Bukowski, was only right regarding stupid people who are full of confidence. Intelligent people equipped with emotional balance/solidity can doubt but not become paralyzed by it, making a decision and pursuing it. Intelligent people full of speculation, deliberation, and self-doubt can't make up their minds. There are also the wise people, a minority among us, who know exactly what to do in this confused world but they are not listened to. This is the problem of the world.

Know your world: Ideals contra reality:

Since the fall of the Berlin Wall and the disintegration of the Soviet Union, most people in the west imagined that now the world will become democratic. In the 2000s, after 9/11, the west that meant that to promote this goal we had to use some force in other countries. In the 2010s, the west began to realize that this method does not work very well, and now, in 2020s, we in the west have given up the idea of a democratic/liberal world and are building walls around us. In 1989 there were around fifteen walls and fences. Today there almost 100 of them.

Self-knowledge:

Human stupidity is visible when people listen to what they wish to hear but not to what they need to know/act upon.

Self-knowledge/knowledge of the world:

If you are sure as hell that your society has a patent on the right way of living a good life in freedom, you are brainwashed.

As long as our brains are designed to glorify ourselves and devaluate others, plus practice shortsightedness, we are in big trouble!

Self-knowledge and knowledge of your world:

"Splitting hairs" or "one cannot see the forest 'cause of the trees" by the mass media and social media has reduced the manifestation of free will among people by creating an unbearable and confusing inferno.

Only a farsighted life view/practice involving free will, impulses, routine, shortsightedness, and day-to-day conformism don't point in this direction.

Strict, rigid faith grants both security and stupidity. Free/critical thinking generates doubt but also access to some free will.

The gravitational forces of society and groupthink cut off whatever wings they grew at a young age for most people.

Petty, envious people won't be able to discern between megalomania on the one hand and megalopsychia on the other hand.

Critical thinking, self-knowledge/knowledge of the human state/the world are preconditions for exercising farsighted free will.

A future where robots and smart machines will do most of our work will be a nightmare for those who will have no meaningful work or task to perform. They will be stamped as useless.

Love, happiness, peace, harmony are extremely overrated values in resolving sapiens' tribulations. Advanced, evolving wisdom is the cure!

Greed, shortsightedness, self-deception, and above all, foolishness are those which bring about destruction in the world!

As long as we don't reduce global population, pollution, production/consumption, we will choke our lives on earth.

There is this huge consumer vanity market selling illusions for people, and lots of human parasites live on it.

There is no happiest country or happiest people. There is only the deluge of "evidence based" nonsense that dumbs people down.

There will not dwell peace on earth as long as we sapiens reign in it. Only by transcending our mental limitations can we hope for peace one day.

Pursuing the ultimate meaning:

The only way to touch eternity is by us ever evolving in order to attain creators stage, dissolving our biological limitations.

What shadowy way in our minds makes the most of us give up the possibility of dancing with stars through evolving hardships?

Knowledge of the world:

The biggest contributor to the infantilization of the masses is mass media/TV coverage constantly shifting our focus.

Take human vanity and mix it with our urge to belong to a group and at the same time be signaled out as unique, and you will know how people are being manipulated to consume excessively.

Soren Kierkegaard, a Danish philosopher, claimed that the deeper you suffer from anxiety, the greater you are (probably thinking of himself, glorifying his misery). Now more than ever we have people suffering from anxiety in Denmark and in the west. I treated many of them, but I must be blind since I could not notice their greatness.

In our western culture, we use both superlatives and lies to describe people as being special and unique, all the while they are just ordinary people. This is counterproductive as It renders people passive.

People obsessed by sex are mentally disturbed, escaping in vain their problems by being intoxicated by it.

Much human mental suffering is caused by modern lifestyles, making humans mental frail/fragile. The cure for the growing mental problems among people is slowing down life's tempo plus upgrading humans beyond their mental frailties and testing their resilience and endurance to challenges.

There are lots humans who act and react as a swarm of flies. Are they real sapiens?

There is nothing impossible for a man who is not forced to follow his words with actions.

It is a frightening defect that humans will not give up their material privileges even when they can see the oncoming deluge.

What is real reality?

Real reality consists of natural laws, where there is a food chain and we are at the top of it and we eat animals and they eat each other, without any consideration of a merciful god. The real reality means human beings are bound by constraints enforced by nature and our brains'capacities, yet with the potentiality to evolve in a wise/sustainable manner beyond them.

Most people have a dubious relationship with hard facts: either they ignore them or bend them as it suits them and their convictions, including their reality. The poet T. S. Eliot said: Humankind cannot bear very much reality. Rejecting reality results in transforming many people into normative psychotics.

The gravitational force of socialization forms most humans into mediocrity/comfort seeking beings. What a loss of potential!

Socrates asked for the point in carrying out a case against a donkey that had kicked him. And truly it is a waste of time, even when the donkey is often a human being. Which reminded me of a story about a rabbi riding together with the owner of the wagon through a muddy path, and realizing that the horse is exhausted. The rabbi gets out of the cart and tries to help the horse. The owner of the cart asks the rabbi why he is helping the horse, and the rabbi retorts, "The animal is exhausted." The owner explains to the rabbi that God created horses in order to carry carts, and rabbis to tell the words of God. The rabbi answers, "I know it, I know it. But when my time comes and I have to report on my deeds down on earth, I don't wish to discuss this matter with a horse."

The mask game:

There are two feelings people cover up even though they are very common. Envy and the feeling of their lives not living up to their expectations/morals.

Think about Davos, where the richest and the mighty meet every year to discuss world problems/challenges. It is self-evident that they play a mask game, as they don't resolve any weighty global problem. The days of the conference are devoted to panel discussions and long meals, and all the problems of the day get discussed, mainly variations of the theme of riding herd on an increasingly fractious world by helping those most in need. And then they party, feel good about themselves for making their little patch-up work. This is what may be considered as a pleasurable and luxurious Potemkin landscape game.

The flagship of democracy, USA, has around 40 million poor people, 70 million obese, 52 million people with emotional problems, more than 40 million who are illiterate, and 2.2 million sitting in prison. A real model to follow.

Cheating due to self-interest in the affairs of man is so massive as to bring regulating efforts on climate in mortal danger.

Self-knowledge:

Genuine self-knowledge implies changes in one's person, actions, and perceptions, which may lead to a sustainable farsighted life view.

You are lucky if you know that you are in spiritual jail. Most people are in a lower jail, without any scent of developed spirituality, and there will they stay to the end of their days. From the spiritual jail, one can escape to freedom, which slowly will turn into a new jail, and so forth.

Human life without struggling to evolve/transcend sustainably is devoid of sustaining and enduring meaning, and therefore is useless for our civilization.

Of course, the human brain possesses some plasticity and our destiny is not totally predetermined. Yet, most people possess brain capacity as to distinguish between good and bad in their human realm, but they lack the capacity to realize that certain contemporary good can turn out to be terrible bad for the future, to-come civilization.

Self-knowledge and knowledge of the world and of our ultimate meaning:

The three fundamental truths that we must face in order to survive/prevail on a long-term basis are as follows:

1. Everything is temporary, including *Homo sapiens*, his achievements, and his constructs.
2. We are, generally speaking, unwise regarding our conduct of the planet's sustainable state. We act, in fact, as its vermin, not as its guardians.
3. Only by transcending ourselves will we pass on our valuable and noble essence to our future upgraded children. Sabotaging this transformation will imply our species' decay and oblivion.

Realizing these truths, what can we do?

We can learn to follow Rumi's words: "Start a huge, foolish project, like Noah . . . It makes absolutely no difference what people think of you."

And that is what need to do, build a new Noah's ark for our future generations

Self-knowledge: Are we self-deceptive mayflies or more?

Oh, Maj fly, made of darkness and light,
frightened by the awaiting eternal night.
You try to keep your horror at bay
by immense self-deceptive might
and a bunch of gods with all might.
As change is our ultimate law/right
with the help of visionary far sight,
our future may become bright,
as we will follow ever-evolving flight.

Knowledge of the world:

Modern civilization faces three immense crises: ecological unsustainability, moral de route, and mental/psychological disturbances.

Democracy and bankruptcy:

The happiest nation in the world, Denmark, is an over consumer society. Already on March 26, 2021, we Danes had consumed the amount of resources that we could rightfully take from the planet in all of 2021. Everything we have then used of resources, we have "stolen" from other nations and the future generations.

If everyone lived on Earth like us Danes, it would require not one but 4.3 globes available to keep balance in the accounts.

Put another way, Denmark does not have the biocapacity to maintain a population like the current one of 5.8 million. In fact, there is only a basis for 1.3 million people on Danish soil with the current level of consumption.

Democracy has run into a decadent stage as its citizens want to save the world without being willing to sacrifice their lifestyle/privileges.

The growing mental misery among youngsters in the west indicates strongly that the west has become decadent. A major reason for the dramatic rise of mentally disturbed/fragile children in the west is the confusion regarding sexual identity.

When castrated teachers teach the students, they will be prone to castrate them too.

Many can't find their place in their own families, can't find their place in society, searching for their "identity" as their lives are without meaningful purpose.

Self-knowledge: Free thinking?

The so-called free thinking of man encounters huge, if not insurmountable, difficulties. The frameworks of our minds are formed to a great degree via conditioning, convictions, and faith. On top of that, many people, if not most, have a hard time controlling their own streams of thought, and become at, least partly, the victims of mental obsessions and compulsions. A third limiting factor of our free thinking is the fact that we are affected by moods, which color our thoughts, sometimes so much as to render it impossible to discern between what is real and what we feel is a reflection of reality. I have come to the conclusion that thinking relatively free implicates proactive and out-of-the-box thinking, bright minds and mental stability, and the nurturing of life perspectives far beyond our collective conditioning.

Wishful thinking is the preferred food of our minds.

Human identity is more based on memories than on personal thoughts and ideas.

Knowledge of the world:

The Anthropocene era started, according to a growing number of geologists, around 1950, five years after my birth. This period was, and is, characterized by human activities that change (for the worse) our world and our life conditions. While more people on earth are richer, more affluent, and self-indulgent, the planet and life on it have suffered tremendously. And there is growing and irrefutable evidence that we are on course toward self-destruction. And this is the roulette we play with our own legacy and the existing life on earth. Some people assume that what we have destroyed we may be able to recreate and heal. But the crucial question is whether this assumption is more wishful thinking than a real option. (Why didn't we stop before the situation got out of our control?)

The alarmingly growing mental misery among youngsters in the west strongly indicates that the west has becomes decadent.

Many can't find their place in their own families; can't find their place in their societies.

Self-knowledge and knowledge of the world and of the ultimate meaning:

People claiming that they never lie are either hypocrites, fools, or psychotic saints, and saints constitute a pommel of this group.

To claim to mitigate the harm done in our global climate debacle sounds to my ears like "a tactical retreat camouflaging defeat," an outright lie.

How are we going to mitigate the ill effects of temperature rising in Israel (2021) by 3.4 °Celsius alone in the last century. Or in the Arctic, 3.1 degrees in the last decades, and in both places it accelerates dramatically.

Know yourself:

Countless people are being nourished by diversions and closed-in circuits of quibbling thoughts without reforming their lives.

Lots of people, when growing old, become sort of spiritual and wish to teach the ignorant. It is rather pathetic sweet nothings. Greatness of mind does not follow the paths of tutored humility/spirituality, resignation, conformity, lust, greed, and pleasure seeking.

One cannot be wise if he/she does not actively support personal and global sustainability, human rights balanced by human obligations, forcefully reducing the economic gap between the rich ones and the poor ones, and our further evolvement.

If a person could live 200 years, was equipped with better brains than the average sapiens, he would consider most of what we utter as superfluous

I have to admit that I hold to a strong conviction that much of our psychological suffering nowadays is the result of too much self-focus, the convenient and mediocre life sold to us as self-realization, and lack of megalopsychia projects in our lives.

Conclusion:

Arthur Schopenhauer said, "The truth passes through three stages. First, it is ridiculed. Second, it is violently opposed. Third, it is accepted as being self-evident." Some people who read my ideas fifteen to twenty years ago ridiculed them. Now my ideas have reached stage two, on the verge of phase three. They are opposed violently, and it is obvious why, as they reflect on the inevitable demise of our civilization and the probable demise of sapiens, and the new upgraded race's rise. This phase of rejection will also pass in due time.

How do I know it?

Because there is no real other viable alternative. All the other alternatives with sapiens as the steering force will be bloodied and end up in our degeneration and decay. Be sure of it!

Footnotes

1) Addiction. Global statistics on addictive behaviors: 2014 status report: Linda R. Gowing, Robert L. Ali, Steve Allsop, John Marsden, Elizabeth E. Turf, Robert West, John Witton First published: 11 May 2015. https://doi.org/10.1111/add.12899

2) Source: Et næsten uladsiggørlig project : John Husted. Weekend Avisen. 23.10.2020 (figures are based on the American bank J. P. Morgan together Vaclav Smil)

3) Michael Hall and Kaitlin Raimi set out to check in a series of experiments in the *Journal of Experimental Social Psychology.*

4) (Mental health - WHO | World Health Organization https://www.who.int › health-topics › mental-health).

5) Addicted To Shopping: 7 Signs You May Have A Problem - WebMDhttps://www.webmd.com › signs-of-a-shopping-addiction)

6) December 1, 2006, Compulsive Buying Disorder Affects 1 in 20 Adults, Causes Marked Distress https://www.psychiatrictimes.com › view › compulsive-bu.

7) https://www.haaretz.co.il/nature/climate/.premium-1.9924032) (https://www.theguardian.com/science/2021/jun/17/earth-trapping-heat-study-nasa-noaa)

8) https://www.haaretz.co.il/nature/climate/.premium-MAGAZINE-1.9903308).

9) Samaneh Ashraf. Ali Nazemi & Amir AghaKouchak. *Scientific Reports* volume 11, Article number: 9135, 2021. Anthropogenic drought dominates groundwater depletion in Iran.

10) Loneliness and Social Isolation Linked to Serious Health Conditionshttps://www.cdc.gov › features › lonely-older-adult

11) 'Generationen der aldrig star op; Peter Hamresen. Weekendavis: Udland.p.8.21.6.2021

12) AntibioticsAutism Spectrum DisorderBrainMicrobiomeNeuroscience PopularRutgers University. By RUTGERS UNIVERSITY, AUGUST 2, 2021

13) *Du kan dø af for meget arbejde*: You can die of over work, May 25, 2021, p. 12

14) Shelly Fan, https://singularityhub.com/2021/06/08/scientists-used-crispr-to-engineer-a-new-superbug-thats-invincible-to-all-viruses/

www.ingramcontent.com/pod-product-compliance
Lightning Source LLC
Chambersburg PA
CBHW021416210526
45463CB00001B/399